Climate Change Adaptation

Nalini Bikkina · Rama Mohana R. Turaga
Editors

Climate Change Adaptation

Traditional Wisdom and Cross-Scale Understanding

Editors
Nalini Bikkina
GITAM School of Humanities
and Social Science
Visakhapatnam, Andhra Pradesh,
India

Rama Mohana R. Turaga
Public Systems Group
Indian Institute of Management
Ahmedabad
Vastrapur, Gujarat, India

ISBN 978-981-97-1075-1   ISBN 978-981-97-1076-8   (eBook)
https://doi.org/10.1007/978-981-97-1076-8

To
Amma & Nanni
With Love

—*Nalini Bikkina*

FOREWORD

"Over the past three months, a heat wave has devastated North India and neighboring Pakistan. Temperatures surpassed 110 degrees Fahrenheit. It is so hot that overheated birds fell out of the sky in Gurgaon, India, and a historic bridge in northern Pakistan collapsed after melting snow and ice at a glacial lake released a torrent of water." (https://www.nytimes.com/2022/06/02/briefing/climate-cha nge-heat-wave-india.html; retrieved on June 24, 2022).

These shocking narratives of the devastating impacts of climate change are increasingly common in both media and public discourses. These heat waves have severely affected farmers' income, impacted food supply, and increased the already growing costs of living for communities across India and across the world. In addition to rising temperatures, climate change has led to frequent floods, soil erosion, and rising sea levels (Dubey et al. 2021). As this excellent compendium that follows this foreword shows, climate hazards have particularly increased the vulnerabilities of coastal communities in India who are often at the frontlines of climate hazards leading to displacements, forced evacuations, and frequent disruptions to the lives of local communities.

While loss of life from climate hazards have been lessened by disaster preparedness through technologies for tracking and warning, such initiatives alone cannot create sustainability among coastline communities. Rebuilding the economies and infrastructure after a disaster requires cooperation and engagement with locals and their traditional systems of indigenous knowledge. Such cooperation will allow policy makers and scientists to leverage indigenous knowledge and anchor their disaster management projects in collective local knowledge systems. It will also require equitable sharing of scientific knowledge through local participation and continued nurturing of "citizen scientists" who may facilitate knowledge sharing between local communities, scientists, and policy makers (Ottinger, 2010; Dawson et al., 2020).

The selection of papers for this compendium eloquently makes the case for leveraging indigenous knowledge and participation in science and policy making for climate change. Several climate concerns are discussed and critically reviewed, including the need to disseminate climate information through storytelling, situating climate discourse within the framework of justice and equity that includes a clear-eyed assessment of class and gendered nature of climate impacts (Rice et al., 2022; Smith et al., 2019; Chu et al., 2018). Others discussed the importance of collection and dissemination of long-term data on climate change and health as well as revisiting the role and impact of the Forest Department in climate discourses. Several papers also identified the need for better information flow between community-driven conservation efforts and scientific and agricultural research as well as the need to listen carefully to grassroots impacts of climate hazards and how locals themselves may identify and adopt the tools needed for building resilient communities.

This is the crux of the matter in debates over building resilient communities imbued with the capacity to absorb ecological disturbances and reorganize itself to its previous state (Popke et al. 2016; Griffin et al. 2017). Yet, effective resilience requires further action and clearer response to critical questions: How can climate policy link resilience with recognition? What steps can be taken to ensure promotion of justice and equity in climate resilience initiatives? And how can we leverage and include lived experiences of communities in building resilient infrastructures? This compendium offers an important roadmap to answer these questions as we seek to build a sustainable future.

Fort Collins, USA   Damayanti Banerjee

REFERENCES

Chu, E., Michael, K. 2018. "Recognition in Urban Climate Justice: Marginality and Exclusion of Migrants in Indian Cities." *Environment and Urbanization.* 31(1): 139–156.

Dawson, T., Hambly, J., Kelley, A., Lees, W., Miller S. 2020. "Coastal heritage, global climate change, public engagement, and citizen science." *Proceedings of the National Academy of Sciences,* 117(15): 8280–8286.

Dubey, A. K., Lal, P., Kumar, P., Kumar, A., & Dvornikov, A. Y. 2021. "Present and future projections of heat wave hazard-risk over India: A regional earth system model assessment." *Environmental Research.* 201: 111573.

Griffin, L., Khalil, D., Allen, A., & Johnson, C. 2017. "Environmental justice and resilience in the urban global South: an emerging agenda." In *Environmental justice and urban resilience in the global South,* Pp. 1–11, Palgrave Macmillan, New York.

Ottinger, G. 2010. "Buckets of resistance: Standards and the effectiveness of citizen science." *Science, Technology, & Human Values,* 35(2): 244–270.

Popke, J., Curtis, S., & Gamble, D. W. 2016. A social justice framing of climate change discourse and policy: Adaptation, resilience and vulnerability in a Jamaican agricultural landscape. *Geoforum,* 73: 70–80.

Smith, J., & Patterson, J. 2019. "Global climate justice activism: "the new protagonists" and their projects for a just transition." In *Ecologically Unequal Exchange.* Pp. 245–272. Palgrave Macmillan, Cham.

Rice, L. J., Long, J., & Levenda, A. 2022. "Against climate apartheid: Confronting the persistent legacies of expendability for climate justice." *Environment and Planning E: Nature and Space,* 5(2): 625–645.

Damayanti Banerjee is a faculty research affiliate at the Department of Sociology at Colorado State University. She received her MA and MPhil in sociology from Jawaharlal Nehru University, India, and her PhD in Environmental Sociology from the University of Wisconsin Madison. Her research and teaching interests are at the intersections of environmental justice, climate change and energy policy, and environmental governance. Dr. Banerjee has conducted in-depth field research on the role of culture and place-making in environmental justice movements, specifically studying how communities' perception of injustices inform key strategies of protests and how successful mobilization strategies are determined by political alliances, social relations, economic preconditions, and cultural contexts.

PREFACE

This book is the documentation of the proceedings of a seminar conducted virtually during 11–15 February 2021. The seminar was organized by GITAM Deemed-to-be University, Visakhapatnam with funding support from the United States-India Educational Foundation (USIEF) through the Fulbright Alumni Award to Nalini Bikkina (the lead editor of this book).

THE SEMINAR

The choice of the theme for the Seminar, Traditional Wisdom and Cross-Scale Understanding, evolved from the lead editor's work with the tribes' people across several locations in India and in the Nebraska state of the United States of America, as part of her Fulbright Academic and Professional Excellence grant. In an effort to connect this work to sustainability (one of the themes of the call for proposals for the Fulbright Alumni Award), the idea of climate change emerged as a relevant area.

The lead editor, in her work with the tribes' people, had studied tribal farming practices, particularly with reference to chemical-free farming that uses traditional practices passed on to them through oral traditions over generations. Considering the complexity of traditional wisdom and its cross-sectionality with the science of climate change, it was felt that a seminar bringing together several voices, stakeholders, and communities onto a single platform would be a suitable format.

The original proposal involved a multi-site format for the seminar, wherein the deliberations were to partly take place on the University campus and partly located in a vulnerable community situated at the foothills of the Eastern Ghats. The idea was to locate the discussion in-situ and to bring in the voices from the community to the deliberations.

However, the COVID-19 pandemic created constraints for the Seminar and did not allow us to take the seminar to the local communities. When USIEF allowed us to do the seminar virtually, our plan was to bring in these local voices live to the session. But then, the communities have expressed difficulties with reference to connectivity.

Despite the difficulties of accessing these community members in person due to COVID protocols, we reached out to a few communities and conducted interviews. We video recorded the interviews and played them in the final sessions of the virtual seminar for discussion.

We rethought the format for the seminar in this context in an attempt to bring diverse, interdisciplinary voices representing multiple stakeholders. We reformatted the deliberations to involve four sessions portraying the work and voices of senior academicians working in the area of climate change in India, young researchers bringing narratives from the field, practitioners and policy makers facilitating adaptation at the community level, and voices from the grassroots. We identified the speakers appropriate for each session by tapping into Fulbright networks, the departments of social sciences and management in Indian universities, civil services, and the Royal Society of Arts (RSA). Thus the selection of the speakers for the sessions was by invitation as opposed to an open call for contributions. The Seminar was scheduled across four sessions—Academic Speak, Research with the Community, Facilitating Adaptation, and From the Grassroots.

In addition to these four sessions, the seminar also featured a keynote speech on the first day by S Palagummi, who is an award-winning journalist, working on a wide range of social issues in the rural parts of India. He has worked extensively in research and documentation of climate change issues at the community level in several parts of India through People's Archive of Rural India (PARI).

THIS BOOK

This book is an effort to disseminate the proceedings of the seminar. Unlike typical conference proceedings, the book is not a collection of a select set of full papers presented at the seminar; it is simply the documentation of stories—unfiltered—that each author brought to the seminar, reflecting on their experiences working in the field.

The process of putting this book together is as follows. We have recorded the online presentations of the authors, transcribed them, and edited them for language and coherence. The initial draft was shared with each author, who further edited the chapters, if necessary, before finalizing them. Communicating climate change using stories and narratives is increasingly considered an effective tool to mobilize action from important stakeholders and in that spirit we hope that the stories in the book will inspire a few. We hope that the book will be of interest to both academics and practitioners alike.

Visakhapatnam, India   Nalini Bikkina
Vastrapur, India   Rama Mohana R. Turaga

Acknowledgments

This volume and the seminar on which it is based are but a new beginning of networks, associations, deep friendships, and a more rigorous academic commitment to climate change adaptation. In the journey of three years working to bring this volume together amidst unprecedented hurdles, we wish to share the credit for the successful completion of the manuscript with a team of people who supported us unconditionally.

We are grateful to the United States-India Educational Foundation for funding the project through the Fulbright Alumni Award. Suranjana Das of USIEF had been a constant go-to person for me in liaising with the USIEF and the Fulbright Association.

At GITAM, we received wholehearted support from the leadership. We are beholden to President Sribharat for his consistent encouragement. He walked the talk by chairing the Keynote Address and had been a pillar of strength and support throughout.

We also take this opportunity to extend our heartfelt gratitude to the authors. What is common to all of them is their deep immersion on the ground. Their humility in accepting to be a part of this modest effort is heartwarming. As each one of them confirmed their participation in the Seminar and the subsequent contribution of chapters to the manuscript, it propelled us to work that much harder to make this manuscript meaningful. Thanks are also due to Prof. Vaibhav Bhamoriya of IIM Kashipur who connected us to several experts in the field.

CONTENTS

Notes on Contributors

Vikram Aditya is Principal Scientist, Wildlife Hunting and Trade Program at the Centre for Wildlife Studies. He was a postdoctoral research associate at the Ashoka Trust for Research in Ecology and the Environment, Bangalore in the National Mission on Biodiversity and Human Well-being. He has a PhD in wildlife conservation biology. His research focused on mammal diversity patterns and landscape change in and around the Papikonda National Park, located in the Northern Eastern Ghats of India. He is currently working on hunting practices of the tribal communities across the Eastern Ghats. He has worked previously with WWF-India and as a 'National Geographic Young Explorer Grantee'.

Nalini Bikkina Currently Professor at GITAM School of Humanities and Social Sciences, Nalini has been an ICSSR Doctoral Fellow. With an interest in Psychology and Public Policy, she worked on interdisciplinary research projects funded by various national agencies like the UGC, ICPR, NCRI, and the Ministry of Education. She was a Fulbright Academic and Professional Excellence Fellow at the University of Nebraska at Omaha, a College of Arts and Letters Fellow at James Madison University, Virginia and a Fellow of the Royal Society of Arts, London. She also received the Fulbright Alumni Award.

Suman Chandra is an Indian Administrative Service Officer and was the district collector of Buldhana, Maharashtra. She has been tendering to the needs of environmental justice among the worst hit rural poor. Besides, she evolved a framework suggestion to the appellate courts for the conservation of the rare Lonar Lake in Buldhana. She led one of the largest livelihood missions of India as the CEO of the Maharashtra Rural Livelihoods Mission. She is also a Limca Book of Records holder for her water conservation campaign in the drought prone district of Osmanabad in Maharashtra. In her ten years in governance and public policy implementation, she has addressed myriad issues related to climate change and has a Master's in Environmental Studies from Yale University through a Fulbright Fellowship.

Carolyn Cobbold is a Research Fellow, investigating food in the nineteenth and twentieth centuries. She completed a history of science PhD at Cambridge University after an early career in journalism. She has also been actively involved in community work, including leading a groundbreaking coastal-planning partnership. Dr. Cobbold was elected a Fellow of the Royal Society of Arts in 2016 for her work on climate change mitigation and community engagement.

Smriti Das is an Associate Professor of Strategic Studies, XLRI Delhi and was earlier in the Department of Policy Studies at TERI-SAS. Her research and scholarship are focused on areas at the interface of environment and development at local and regional scale. She specifically engages with topics such as environmental policy, processes and politics, forest policy and governance, institutional analysis, sustainable livelihoods and communities, decentralized governance, gender, and resource politics. She is working on unpacking the concept of community in forest governance; mainstreaming climate change in local governance and planning.

Satya Kishan Kumar Namala Currently a Doctoral Fellow in Development Studies from GITAM School of Humanities and Social Sciences, his research interests focus on Climate Change Adaptation and Resilience. He completed his Masters in Development Studies from the Indian Institute of Technology, Mandi, and was an Academic and Research Assistant at the Indian Institute of Management, Visakhapatnam.

Haritha S. Narayanan served as an Academic Associate within the Public Systems Group at the Indian Institute of Management, Ahmedabad. She completed her postgraduate studies in Politics and International Relations from Pondicherry University. Her research interests encompass climate governance, community studies, and human rights. She worked as a research assistant in various institutions, including the Manohar Parrikar Institute for Defence Studies and Analyses, the Climate and Energy Policy Research Lab and the Indian Institute of Technology, Kanpur, focusing on pertinent areas of public policy.

Sainath Palagummi is an Indian journalist who focuses on social and economic inequality, rural affairs, poverty, and the aftermath of globalization in India. He is the founder and editor of the People's Archive of Rural India (PARI). He was the Rural Affairs Editor at The Hindu. Amartya Sen has called him "one of the world's great experts on famine and hunger." He was conferred an Honorary Doctor of Letters degree by the University of Alberta, Edmonton. He is one of the few Indians to receive the Ramon Magsaysay Award, Asia's most prestigious award, in 2007, for his "passionate commitment as a journalist to restore the rural poor to India's national consciousness." He was the first Indian to win the Magsaysay in the category of Journalism, Literature and Creative Communications Arts in nearly 25 years after R. K. Laxman. He was the McGraw Professor of writing at Princeton University, 2012. He was also the first reporter in the world to win Amnesty International's Global Human Rights Journalism Prize in its inaugural year in 2000. He also won the United Nation's Food and Agriculture Organization's Boerma Prize in 2000, the Harry Chapin Media Award in New York, 2006, and the European Commission's Lorenzo Natali Prize in 1995. In 1984 he was a Distinguished International Scholar at the University of Western Ontario and in 1988 a visiting lecturer at Moscow University. He was also a Distinguished International Professional at Iowa University (Fall 1998), the first McGill Fellow at Trinity College, Hartford, Connecticut (Spring 2002), and Visiting Professor at University of California, Berkeley (Fall 2008).

A. Rama Mohana Reddy is an Indian Forest Service Officer with 32 years of experience in afforestation bridging the government and the communities, especially in the Himalayan region, along with stabilization of Himalayan riverbanks including Ravi, Chenab, Beas, and Yamuna. He was involved in approvals of forestlands for development projects. He had extensively trained Indian Forest Service Officers and forest communities in forest management with special focus on GIS.

Sunil D. Santha As an academic with keen interests in the field of environmental risks, climate justice, and livelihood uncertainties, he strives towards understanding the role of social institutions and participatory action in reducing vulnerabilities and strengthening just adaptation practices. He believes in action research towards innovating participatory methods of entrepreneurial action and emergent livelihoods.

Bijayashree Satpathy holds a doctorate from Tata Institute of Social Sciences, Mumbai. Her research is on natural resources governance. Before working as a full-time researcher, she worked with various national and international non-profit organizations that include Centro Internacional de Mejoramiento de Maíz y Trigo (CIMMYT)—a non-profit agricultural research and training institute. She has extensive fieldwork experiences in the regressive regions of India, and has been a post-doctoral research fellow at the Graduate Institute of International and Development Studies, Geneva.

Priya Tallam Conservationist Designer and Socio-Biologist. Priya Tallam worked as an architect in India on historic preservation, and an urban and regional planner in CA, USA for city and county government. With a deep interest in data analytics, she took on GIS for public service (public safety, planning, public health, habitat conservation, and violence prevention) and specifically, water conservation. Traveling in tropically rich India guided her next step: chairing a non-profit for the cause of biodiversity sustenance. She is working with VSPCA, the Visakha Society for the Protection and Care of Animals, expanding their strategic and tactical plans, and augmenting science and technology with cultural knowledge (and knowledge from Mother Nature). She is studying the political ecology and economic and cultural geography of the city, in an effort to foster critical thinking between scientists, planners, and thinkers, and spur the city's residents to grassroots action to preserve their bio-diverse heritage keeping Visakhapatnam forever a biophilic city.

Marcus Taylor is an Associate Professor and Head of the Department of Global Development Studies at Queen's University, Kingston, Canada. He researches and teaches agriculture, labor, and livelihoods. His recent books include *The Political Ecology of Climate Change Adaptation* (Routledge 2015) and *Global Labour Studies* (with Sébastien Rioux, Polity Press, 2018). He is currently completing a volume on *Climate Smart Agriculture: A Critical Perspective* that is based in part on case studies of agricultural intensification and climate-resilient agricultural projects in Telangana and Karnataka. He is a contributing author to the forthcoming IPCC AR6 Working Group 2 chapter on climate-resilient development.

Rama Mohana R. Turaga is currently a faculty with the Public Systems Group at the Indian Institute of Management Ahmedabad (IIMA). He was a Research Associate at Dartmouth College, Hanover and a Postdoctoral Research Fellow at the Georgia Institute of Technology, from where he was also awarded a PhD in Public Policy. His research interests are broadly in sustainability governance, with a specific focus on environmental regulations, business sustainability, and collective action for sustainability. At IIMA, he teaches courses on sustainability governance and public policy.

LIST OF FIGURES

Leveraging Human–Natural World Intersections for Climate Change Adaptation

Listening to Locals

Introduction

Rama Mohana R. Turaga, Nalini Bikkina,
and Haritha S. Narayanan

Abstract Climate change presents an unprecedented challenge for human existence on earth. While carbon mitigation has been the focus thus far, it is now well recognized that the adaptation to already visible impacts of the historical emissions of carbon needs serious attention. Mitigation requires global collective action; adaptation must factor in local ecological and social contexts. Therefore, local and traditional knowledge is significant in planning for adaptation. This chapter reviews the literature on the role of traditional and local knowledge in natural resources management, with a specific focus on climate adaptation. It is apparent from the review that integrating western scientific knowledge with traditional and local knowledge, sometimes termed co-production, has become an emerging theme in both academic literature and practice. In addition, the chapter outlines the manner in which local and traditional knowledge could find a place in India's climate adaptation policies. The chapter concludes with a summary of the contributions in this volume.

R. M. R. Turaga (✉) · H. S. Narayanan
Indian Institute of Management Ahmedabad, Ahmedabad, India
e-mail: mohant@iima.ac.in

N. Bikkina
GITAM School of Humanities and Social Sciences, Visakhapatnam, India

Keywords Climate change · Adaptation · Indigenous knowledge · Co-production · Conservation

It is now indisputable that climate change is an unprecedented challenge confronting humanity. While the world is scrambling to mitigate the greenhouse gas emissions to limit the global average temperature rise, there is also an acute realization that the historically emitted carbon is already causing serious impacts to which we should adapt. Thus in the last decade or so, climate adaptation has emerged as the second pillar to mitigation in climate action (Orlove, 2022). Although the definitions of adaptation might vary, one that encompasses many common elements defines climate adaptation as "policies, proactive or reactive, that seek to reduce the biophysical, social, and economic vulnerability (or enhance resilience) of a given area, organization, population group, or individuals to climate change" (Dolšak & Prakash, 2018, p. 319).

Carbon mitigation requires global collective action in which national governments agree to cut carbon emissions to keep the global average temperature rise below a certain target level. Adaptation actions, on the other hand, often must take place at the subnational and local scales. Adaptation at the local level may include, for example, hard infrastructure (e.g., sea walls, rain water harvesting structures) and soft infrastructure (e.g., community capacity building). Regardless of the type of adaptation measures, effective adaptation planning requires consideration of local ecological and social context. Thus, there has been an increasing recognition of the need to account for local and traditional knowledge in the adaptation planning process.

TRADITIONAL AND LOCAL KNOWLEDGE: CONCEPTUAL DEFINITIONS AND BACKGROUND

The academic literature and the international practice use multiple terms, often interchangeably, to refer to traditional and/or local knowledge systems. Such terms include indigenous knowledge (IK), local knowledge, traditional ecological knowledge (TEK), indigenous and local knowledge (ILK), and so on. The Intergovernmental Platform on Biodiversity and Ecosystems Services (IPBES) distinguishes between indigenous people and local communities:

"Indigenous peoples include communities, tribal groups and nations, who self-identify as indigenous to the territories they occupy, and whose organization is based fully or partially on their own customs, traditions, and laws (Hill et al., 2020, p. 9)" whereas "Local communities are groups of people who maintain intergenerational connection to place and nature through livelihood, cultural identity, worldviews, institutions and ecological knowledge. (Hill et al., 2020, p. 9)"

The IPBES defines the ILK systems as "...bodies of integrated, holistic, social and ecological knowledge, practices and beliefs pertaining to the relationship of living beings, including people, with one another and with their environments" (Hill et al., 2020, p. 11). Rarai et al., (2022) provide a more elaborate conceptualization of ILK systems: "...a place-specific nested knowledge system of information-practises-values that consists of three key dimensions" (p. 2243). The three dimensions include (i) information about the environment (e.g., biophysical conditions, seasonal indicators, classification systems for species), (2) governance and management arrangements for natural resources use and addressing environmental risk, and (3) worldviews (e.g., values and ethics), which shape people's understanding of the relationship between humans and the natural environment. ILK systems are highly location-specific, collectively held, and are typically passed on from one generation to the other orally or through practice (Rarai et al., 2022). On similar lines, Menzies and Butler (2006) outline cumulative and long-term, dynamic, historical, local, holistic, embedded, and moral and spiritual as the key attributes of most definitions of traditional ecological knowledge (TEK) systems.

The recognition of ILK systems in natural resource management precedes the rise of climate change as a major environmental concern. The Convention on Biodiversity, which came into effect in 1993, recognized the rights of indigenous people and the importance of their knowledge systems in conserving biodiversity (Mauro & Hardison, 2000). The 1992 Rio Declaration and the Intergovernmental Forum on Forests are the other examples of international forums that acknowledged the rights of indigenous population and their knowledge systems. In the climate change context, responding to the concerns of inadequate representation of the views of indigenous and traditional knowledge systems (Orlove, 2022), the UN Framework Convention on Climate Change (UNFCCC) had set up a "Local Communities and Indigenous Peoples Platform" in

2017 to ensure better representation of indigenous and local communities in UNFCCC processes (Shawoo & Thornton, 2019).

TRADITIONAL AND LOCAL KNOWLEDGE VS. SCIENTIFIC KNOWLEDGE: A CASE FOR INTEGRATION?

One of the key advantages of ILK systems is the rich local knowledge that they can provide to the understanding of natural ecosystem processes (DeWalt, 1994). The traditional knowledge systems are based on observations over a long time scale (Clarke, 1990) because of which they represent the "cumulative and dynamic product of many generations of experience and practice" (Menzies & Butler, 2006, p. 2). Some scholars argue that the emergence of ecosystem-based management, which incorporates system-level understanding of ecological processes has drawn greater attention to the ILK systems (e.g., Menzies & Butler, 2006). The dependence of indigenous and local communities on the local natural resources for their livelihoods requires them to develop an intimate understanding of the interrelationships among various parts of the local ecosystems. These characteristics—information on the local environment over long time-scales and the intimate system-level knowledge—make traditional knowledge valuable in the context of climate change adaptation. This enables the ILK systems to "develop the sensitivity to 'read' critical signs and signals that something unusual is happening" (Berkes, 2009, p.153).

The value of ILK systems in natural resource management generally and climate change adaptation in particular are well documented (e.g., Berkes, 2009; Clarke, 1990; Galappaththi et al., 2019; Granderson, 2017; Leonard et al., 2013; Mardero et al., 2023; Menzies, 2006; Nkuba et al., 2020). Given this, a key question has been to understand the ways in which the knowledge from ILK systems can be integrated into the western scientific knowledge systems. Early work pitted the two knowledge systems against each other (see DeWalt, 1994). Those supporting the western scientific method argued that traditional indigenous knowledge systems are irrelevant in the context of the vast technological progress that characterize modern natural resource extraction and management, including in forestry, agriculture, fisheries, and so on. The supporters of ILK systems pointed to the environmentally harmful and socially unjust outcomes of modern scientific methods. Over time, this debate evolved to suggest that it is possible to identify synergies between the two forms of knowledge—despite their differing ontologies and epistemologies—and integrate them to generate more effective and sustainable natural resource management policies (e.g., Becker & Ghimire, 2003; Berkes, 2009; Rarai et al., 2022). In addition to the instrumental value of improving resource management and climate adaptation planning, normatively, the lack of integration of the ILK systems constitute "epistemic injustice" (Byskov & Hyams, 2022) on the grounds of inadequate representation of indigenous and local communities, who are also disproportionately more vulnerable to climate change.

The integration of western scientific knowledge with ILK systems has become a dominant theme in the literature and is often termed as "co-production" (e.g., Berkes, 2009). For example, the journal *Ecology & Society* published a special issue in September 2013 titled "Integrating Indigenous Ecological Knowledge and Science in Natural Resource Management: Perspectives from Australia" with a focus on co-production in Australia. Similarly, Gianelli et al., (2021) analyze the attempts to integrate traditional knowledge and scientific knowledge in the management of small-scale fisheries across the world. In one of the more recent studies, van Maurik Matuk et al. (2023) document and analyze 13 case studies of natural resource management that involve co-production by simultaneously incorporating scientific knowledge systems and ILK systems.

While traditional knowledge is widely acknowledged as critical in natural resource management, serious concerns have been raised regarding the processes of integration. A key theme in this literature is the vast power asymmetries between western scientific experts and indigenous and local communities and its impact on indigenous and local people (e.g., Bohensky et al., 2013; Shawoo & Thornton, 2019; Thompson et al., 2020). The privileging of scientific knowledge as objective, rigorous, and universal as opposed to the indigenous knowledge as subjective, arbitrary, and local (Mistry & Berardi, 2016) creates unequal representation in the integration process (Manrique et al., 2018). Further, as Rarai et al., (2022) show in Vanuatu, the domination of external scientific expertise in local adaptation planning could lead to adaptation policies that worsen the vulnerability of indigenous and local communities to climate change. This is because of the replacement of the adaptation practices derived from ILK systems with the adaptation "solutions" prescribed by the western scientific knowledge system, which operates in a distinctly different ontological and epistemological regime (Mistry & Berardi, 2016; Rarai et al., 2022).

The question of how to meaningfully bring ILK systems to bear upon more effective and just natural resources management and climate adaptation thus requires bridging the power asymmetries. For example, in their work on ILK approach within IPBES, scholars identified the challenges, prevailing practices that worked, and the remaining gaps for a more effective ILK approach (Hill et al., 2020). The four classes of prevailing practices in their approach include: (i) respecting the rights of indigenous people and local communities (IPLC), (ii) supporting care and mutuality, (iii) strengthening IPLC and their knowledge systems, and (iv) supporting knowledge exchange. Insights from such approaches could serve as a starting point for a more just representation of ILK systems in climate adaptation. Other scholars suggest approaches such as "co-existence" (as opposed to co-production) (Nursey-Bray et al., 2020) and indigenous articulations (Diver, 2017) as means to overcome the power asymmetry.

Traditional and Local Knowledge: The Indian Context

The role of local and traditional knowledge in Indian environmental conservation—for example, biodiversity conservation, agriculture, and forestry—has been well documented. Madhav Gadgil's work documented case studies of local conservation efforts and advocated for greater local involvement in biodiversity conservation (e.g., Gadgil, 1992, 1995; Gadgil et al., 1993, 2003). A unique initiative in the Indian context with regard to the documentation of traditional knowledge is the People's Biodiversity Register (PBR) program, initiated by a network of nongovernmental organizations (NGOs) in the mid-1990s. The PBR program "is an attempt to record people's knowledge and perceptions of the status, uses, history, ongoing changes and forces driving these changes in the biological diversity resources of their own localities" (Gadgil, 2000, p. 2).

Another stream of literature in the Indian context documents the role of traditional and local knowledge with specific geographical focus. Singh et al. (2010) analyze efforts to learn and document the traditional conservation practices among three tribal communities in northeastern India. Rao and Ramana (2007) examine the traditional knowledge and resource management practices of primitive tribes in Andhra Pradesh in southern India, and Chaudhry et al., (2011) describe the sustainable agricultural practices of the Apatani tribe in Arunachal Pradesh. The literature on local and traditional knowledge is, however, relatively sparse in the context of climate adaptation (see Coulthard, 2008; Lebel, 2013; Shimrah, 2017; Singh et al., 2012, for some examples of community adaptation to climate change using local knowledge and practices).

India's Climate Policy and Local Participation

India, with its long coastline and high population density, ranks among the countries most at risk from climate change. In recognition of both its responsibilities as an emerging economy with increasing contribution to greenhouse gas emissions and the potential impacts of climate change on its population, India launched a comprehensive National Action Plan on Climate Change (NAPCC) in 2008. The NAPCC, in conjunction with the State Action Plans on Climate Change (SAPCCs) and the Nationally Determined Contribution (NDC) under Paris Agreement provide a comprehensive framework for climate mitigation and adaptation action at the federal and state levels. India has also set up in 2015, a National Adaptation Fund on Climate Change (NAFCC), a federal government initiative to support climate change adaptation. This federal grant facilitates the objectives of the NAPCC and supports operationalization of SAPCCs. The Fund supports adaptation projects in pivotal sectors enlisted in the SAPCCs that are not funded by any ongoing central and state government schemes (Prasad & Sud, 2019). Despite these initiatives, climate adaptation action in India is largely considered inadequate (see Patra, 2016). Funding support is extremely inadequate even when the pledged support is secured (Flam & Skjaerseth, 2009; Hochrainer-Stigler et al., 2014; Smith et al., 2011). For example, in 2019–2020, India spent only 8% of the expenditures required for adaptation (Singh, 2023).

Much of India's climate adaptation policy is top-down. The local nature of climate change adaptation has, however, been acknowledged in the SAPCCs, which adopted a consultative process with stakeholders and published state specific climate change adaptation priorities. The SAPCCs factored in the conception that adaptation is a locally situated occurrence and therefore the local communities have a crucial role to play in the success of these interventions (Barret, 2013; Remling & Persson, 2015).

Evidence, however, shows great variation across states in ensuring participation and the final plans rarely reflect the deliberations of the consultative process (e.g., Dubash & Jogesh, 2014; Jogesh & Dubash, 2015). Scholars call for institutional processes that enable greater local participation in adaptation planning and implementation to take advantage of traditional and local knowledge systems, deal with uncertainty, and also develop more locally-sensitive and just policies (e.g., Dhanapal, 2014; Kodiveri & Sen, 2022; Mehta et al., 2019; Singh, 2023). This volume contributes to these conversations by outlining several case studies of the knowledge systems and competencies of local communities to contribute to climate change adaptation and natural resource management.

Contributions of this Volume

In Chapter 2, noted environmental journalist P Sainath documents the adverse consequences of climate change across a spectrum of rural areas in India and highlights the resilience and ingenuity of the local communities in coming up with innovative ideas that help cope with climate change. He discusses a "climate apartheid" scenario that contributes to a growing inequality and increasing levels of deprivation among the vulnerable sections of the community. He believes that climate change is absent in the public discourse and calls for a different kind of reporting through the voices and lived experiences of the vulnerable communities.

Sunil Damodaran (Chapter 3) focuses on the moral connotations of climate change adaptation strategies through the life stories of small fishing communities in India. The processes and consequent outcomes of adaptation strategies need to factor in equality, justice, care, and empowerment of the disadvantaged sections of society who are highly vulnerable to the impacts of climate change. The chapter points to the need for adaptation knowledge, actions, and movements to be situated from the standpoints of the marginalized actors.

Marcus Taylor argues in Chapter 4 that the Indian agricultural extension system has an uncomfortable relationship with the traditional knowledge of the farmer. The system pursues a top-down technology transfer approach and does not foreground local knowledge including cultivation practices that minimize external inputs. He argues that farmer knowledge is not restricted to cropping choices and cultivation patterns but also includes an understanding of how to manage the accompanying socio-cultural relationships that are an overarching dimension of agriculture-based livelihoods. Climate change resilience, particularly in the rural areas can be facilitated and strengthened by expanding the adaptation framework to include farmer knowledge.

The role of women in local knowledge systems is one of the under-studied aspects. In Chapter 5, Smriti Das attempts to mainstream communities and commons into debate on climate change from a gendered perspective. She acknowledges that the impacts of climate change are not similar cross-sectionally and that some segments of the population will be affected more than the others. She discusses marginalities that limit the adaptive capacities and unpacks gendered vulnerabilities to climate change and establishes the need for more gender-responsive policies and institutional designs to help women adapt to changing climatic conditions.

In Chapter 6, Vikram Aditya shows in his empirical work that the forest-dependent communities of the northern Eastern Ghats perceived significant forest degradation owing to climate change, plantation farming, and dam building. In planning the construction of large dams, it is imperative to include the perspective of the local communities and factor in their struggles at adaptation to human intervention in their natural habitats and the adversities of climate change that these communities are trying to adapt to.

Chapter 7 by Bijayashree Satpathy presents an instance of in situ conservation of forests through maintenance and management of sacred groves by the forest dwelling communities. These sacred groves preserve the pristine forests and are reservoirs of traditional knowledge for the tribal communities. A robust and effective implementation of the Forest Rights Act protecting the rights of these communities over the sacred groves will help these communities adapt to the changing climate.

Suman Chandra focuses on climate adaptation in cities in Chapter 8. In stating that urbanization is irreversible, Suman advocates decarbonization pathways and responsible energy consumption by the cities. Her thesis is that the cities are more efficient than the villages with reference to economies of scale and therefore planned urbanization is an effective climate adaptation measure. She argues for the identification of climate goals for every subnational entity through the lens of carbon-negative cities and villages through bottom-up approaches which need to be a healthy blend of traditional and modern methods.

A Rama Mohan Reddy (Chapter 9) brings his experience as a forest service bureaucrat in articulating how the state agencies facilitate coping mechanisms for traditional forest dwellers. He advocates for a net-zero deforestation policy to mitigate climate change and enhance biodiversity conservation and envisages a crucial role for forest bureaucracy in buffering the impact of climate change on forest dwelling communities by training them and co-opting them to support afforestation as an adaptation mechanism.

Chapters 10 and 11 by Priya Tallam and Carolyn Cobbold respectively are based on their immersive experience in local collaborative efforts in natural resource management. Priya highlights the role of local organizations and their collaboration with state agencies in developing location-specific strategies for protecting biodiversity threatened by climate change. She shares her experience working for Visakha Society for the Protection and Care of Animals (VSPCA) with the state department of environment, Andhra Pradesh to protect olive ridley turtles, increasingly threatened by local development and climate change, on the coast of Visakhapatnam. In particular, she describes the transformation in the role of dogs from being predators of turtles to collaborators in their protection through local collaboration and management.

Carolyn Cobbold through her work on the Manhood Peninsula in Britain showcases the potential impact of collaborative networks at the local level in understanding the impacts of climate change and in collectively conceptualizing and executing integrated and sustainable solutions. A managed coastal realignment scheme as a sustainable sea defense has helped channelize the energy of the sea and helped create several hundred acres of compensatory wetland. The consequent changes in the drainage pattern improved the wildlife in the area. Carolyn argues that local communities working together outside of political, administrative, and disciplinary boundaries can bring in innovative solutions that help facilitate adaptation to climate change.

Finally, in the last Chapter, Namala Satya Kishan Kumar and Nalini Bikkina discuss the manifestations and impacts of climate change as narrated by fisherfolk and tribal farmers along the coastal area of Visakhapatnam in India using the public adaptation stages framework. The chapter presents the argument that it is imperative to critically explore the diverse climate change challenges for indigenous communities and the structural roadblocks that restrict the capacity of especially the marginalized people to adapt to climate change, more so in times of resource insufficiency. The vulnerable communities in turn essentially locate their adaptation efforts within the framework of transformative adaptation, which attempts to modify the basic attributes of the existing systems as a response to the actual or anticipated effects of change.

REFERENCES

Barrett, S. (2013). Local level climate justice? Adaptation finance and vulnerability reduction. *Global Environmental Change—Human and Policy Dimensions, 23*, 1819–1829.

Becker, C. D., & Ghimire, K. (2003). Synergy between traditional ecological knowledge and conservation science supports forest preservation in Ecuador. *Conservation Ecology, 8*(1), 1.

Berkes, F. (2009). Indigenous ways of knowing and the study of environmental change. *Journal of the Royal Society of New Zealand, 39*(4), 151–156.

Bohensky, E. L., Butler, J. R., & Davies, J. (2013). Integrating indigenous ecological knowledge and science in natural resource management: Perspectives from Australia. *Ecology and Society, 18*(3), 20.

Byskov, M. F., & Hyams, K. (2022). Epistemic injustice in climate adaptation. *Ethical Theory and Moral Practice, 25*(4), 613–634.

Chaudhry, P., Dollo, M., Bagra, K., & Yakang, B. (2011). Traditional biodiversity conservation and natural resource management system of some tribes of Arunachal Pradesh, India. *Interdisciplinary Environmental Review, 12*(4), 338–348.

Clarke, W. C. (1990). Learning from the past: Traditional knowledge and sustainable development. *The Contemporary Pacific, 2*(2), 233–253.

Coulthard, S. (2008). Adapting to environmental change in artisanal fisheries—Insights from a South Indian Lagoon. *Global Environmental Change, 18*(3), 479–489.

DeWalt, B. (1994). Using indigenous knowledge to improve agriculture and natural resource management. *Human Organization, 53*(2), 123–131.

Dhanapal, G. (2014). Climate adaptation in India. *Nature Climate Change, 4*(4), 232–233.

Diver, S. (2017). Negotiating Indigenous knowledge at the science-policy interface: Insights from the Xáxli'p Community Forest. *Environmental Science & Policy, 73*, 1–11.

Dolšak, N., & Prakash, A. (2018). The politics of climate change adaptation. *Annual Review of Environment and Resources, 43*, 317–341.

Dubash, N. K., & Jogesh, A. (2014). From margins to mainstream? State climate change planning in India. *Economic and Political Weekly, 49*(48), 86–95.

Flam, K. H., & Skjaerseth, J. B. (2009). Does adequate financing exist for adaptation in developing countries? *Climate Policy, 9*, 109–114.

Gadgil, M. (1992). Conserving biodiversity as if people matter: A case study from India. *Ambio, 21*(3), 266–270.

Gadgil, M., Berkes, F., & Folke, C. (1993). Indigenous knowledge for biodiversity conservation. *Ambio, 22*(2/3), 151–156.

Gadgil, M. (1995). Reckoning with life. *The Administrator, 40*(1), 93–101.

Gadgil, M. (2000). People's biodiversity registers: Lessons learnt. *Environment, Development and Sustainability, 2*, 323–332.

Gadgil, M., Olsson, P., Berkes, F., & Folke, C. (2003). Exploring the role of local ecological knowledge in ecosystem management: three case studies. In. F. Berkes., J. Colding, & C. Folke (Eds.), *Navigating social-ecological systems: Building resilience for complexity and change* (pp. 189–209). Cambridge University Press.

Galappaththi, E. K., Ford, J. D., Bennett, E. M., & Berkes, F. (2019). Climate change and community fisheries in the Arctic: A case study from Pangnirtung, Canada. *Journal of Environmental Management, 250*, 109534.

Gianelli, I., Ortega, L., Pittman, J., Vasconcellos, M., & Defeo, O. (2021). Harnessing scientific and local knowledge to face climate change in small-scale fisheries. *Global Environmental Change, 68*, 102253.

Granderson, A. A. (2017). The role of traditional knowledge in building adaptive capacity for climate change: Perspectives from Vanuatu. *Weather, Climate, and Society, 9*(3), 545–561.

Hill, R., Adem, Ç., Alangui, W. V., Molnár, Z., Aumeeruddy-Thomas, Y., Bridgewater, P., & Xue, D. (2020). Working with indigenous, local and scientific knowledge in assessments of nature and nature's linkages with people. *Current Opinion in Environmental Sustainability, 43*, 8–20.

Hochrainer-Stigler, S., Mechler, R., Pflug, G. & Williges, K. (2014). Funding public adaptation to climate related disasters. Estimates for a global fund. *Global Environmental Change—Human and Policy Dimensions, 25*, 87–96.

Jogesh, A., & Dubash, N. K. (2015). State-led experimentation or centrally-motivated replication? A study of state action plans on climate change in India. *Journal of Integrative Environmental Sciences, 12*(4), 247–266.

Kodiveri, A., & Sen, R. (2022). Climate action plans and justice in India. In. P. Kashwan. (Ed). *Climate Justice in India: Volume 1* (pp. 115–139). Cambridge University Press.

Lebel, L. (2013). Local knowledge and adaptation to climate change in natural resource-based societies of the Asia-Pacific. *Mitigation and Adaptation Strategies for Global Change, 18*, 1057–1076.

Leonard, S., Parsons, M., Olawsky, K., & Kofod, F. (2013). The role of culture and traditional knowledge in climate change adaptation: Insights from East Kimberley, Australia. *Global Environmental Change, 23*(3), 623–632.

Manrique, D. R., Corral, S., & Pereira, Â. G. (2018). Climate-related displacements of coastal communities in the Arctic: Engaging traditional knowledge in adaptation strategies and policies. *Environmental Science & Policy, 85*, 90–100.

Mardero, S., Schmook, B., Calmé, S., White, R. M., Chang, J. C. J., Casanova, G., & Castelar, J. (2023). Traditional knowledge for climate change adaptation in Mesoamerica: A systematic review. *Social Sciences & Humanities Open, 7*(1), 100473.

Mauro, F., & Hardison, P. D. (2000). Traditional knowledge of indigenous and local communities: International debate and policy initiatives. *Ecological Applications, 10*(5), 1263–1269.

Mehta, L., Srivastava, S., Adam, H. N., Alankar, B. S., Ghosh, U., & Kumar, V. V. (2019). Climate change and uncertainty from 'above' and 'below': Perspectives from India. *Regional Environmental Change, 19*, 1533–1547.

Menzies, C. R., & Butler, C. (2006). Introduction: Understanding ecological knowledge. In C. R. Menzies (Ed.), *Traditional ecological knowledge and natural resource management* (pp. 1–17). University of Nebraska Press.

Menzies, C. R. (2006). *Traditional ecological knowledge and natural resource management*. University of Nebraska Press.

Mistry, J., & Berardi, A. (2016). Bridging indigenous and scientific knowledge. *Science, 352*(6291), 1274–1275.

Nkuba, M. R., Chanda, R., Mmopelwa, G., Kato, E., Mangheni, M. N., & Lesolle, D. (2020). Influence of indigenous knowledge and scientific climate forecasts on arable farmers' climate adaptation methods in the Rwenzori region, Western Uganda. *Environmental Management, 65*, 500–516.

Nursey-Bray, M., Palmer, R., Stuart, A., Arbon, V., & Rigney, L. I. (2020). Scale, colonisation and adapting to climate change: Insights from the Arabana people, South Australia. *Geoforum, 114*, 138–150.

Orlove, B. (2022). The concept of adaptation. *Annual Review of Environment and Resources, 47*(1), 535–581.

Patra, J. (2016). Review of current and planned adaptation action in India. CARIIA Working Paper # 10. Retrieved December 3, 2023, from https://idl-bnc-idrc.dspacedirect.org/server/api/core/bitstreams/46d699c8-3844-49fa-8adc-307b9caaa7bd/content

Prasad, R. S., & Sud, R. (2019). Implementing climate change adaptation: Lessons from India's National Adaptation Fund on Climate Change (NAFCC). *Climate Policy, 19*(3), 354–366.

Rao, V. L. N., & Ramana, G. V. (2007). Indigenous knowledge, conservation and management of natural resources among primitive tribal groups of Andhra Pradesh. *Anthropologist, 3*, 129–134.

Rarai, A., Parsons, M., Nursey-Bray, M., & Crease, R. (2022). Situating climate change adaptation within plural worlds: The role of Indigenous and local knowledge in Pentecost Island, Vanuatu. *Environment and Planning e: Nature and Space, 5*(4), 2240–2282.

Remling, E., & Persson, A. (2015). Who is adaptation for? Vulnerability and adaptation benefits in proposals approved by the UNFCCC adaptation fund. *Climate and Development, 7*, 16–34.

Shawoo, Z., & Thornton, T. F. (2019). The UN local communities and Indigenous peoples' platform: A traditional ecological knowledge-based evaluation. *Wiley Interdisciplinary Reviews: Climate Change, 10*(3), e575.

Shimrah, T. (2017). Traditional ecological knowledge system as climate change adaptation strategies for mountain community of Tangkhul Tribe in Northeast India. *International Journal of Environmental and Ecological Engineering, 11*(7), 686–693.

Singh, R. K., Pretty, J., & Pilgrim, S. (2010). Traditional knowledge and biocultural diversity: Learning from tribal communities for sustainable development in northeast India. *Journal of Environmental Planning and Management, 53*(4), 511–533.

Singh, R. K., Turner, N. J., & Pandey, C. B. (2012). "Tinni" Rice (*Oryza rufipogon* Griff.) production: An integrated sociocultural agroecosystem in Eastern Uttar Pradesh of India. *Environmental Management, 49*, 26–43.

Singh, K. (2023, July 27). Climate change adaptation is the need of the hour. *India Development Review*. Retrieved December 3, 2023, from https://idronline.org/article/climate-emergency/climate-change-adaptation-is-the-need-of-the-hour/

Smith, J. B., Dickinson, T., Donahue, J. D. B., Burton, I., Haites, E., Klein, R. J. T., & Patwardhan, A. (2011). Development and climate change adaptation funding: Coordination and integration. *Climate Policy, 11*, 987–1000.

Thompson, K. L., Lantz, T., & Ban, N. (2020). A review of Indigenous knowledge and participation in environmental monitoring. *Ecology and Society, 25*(2), 10.

van Maurik Matuk, F. A., Verschuuren, B., Morseletto, P., Krause, T., Ludwig, D., Cooke, S. J., & dos Reis Carvalho, B. L. (2023). Advancing co-production for transformative change by synthesizing guidance from case studies on the sustainable management and governance of natural resources. *Environmental Science & Policy, 149*, 103574.

Learning Climate Change Adaptation from Traditional Communities

Sainath Palagummi

Abstract Serious consideration of Climate Change and its attendant devastation is most absent in public discourse in this country. The media is either unable to tell the story or tell it in a way that is understandable to ordinary citizens. This is partly because the media are entrenched within the economic and social systems that are significant drivers of climate change. Further, reporting on climate change largely consists of covering inter-governmental conferences, negotiations, and very dense scientific reports that, important as they may be, are incomprehensible to lay readers and viewers. This talk will look at how this calls for a very different kind of reporting—climate change through the voices and lived experiences of everyday people.

Keywords Climate change · Public discourse · Climate apartheid · Agrarian crisis · Lived experiences

S. Palagummi (✉)
People's Archive of Rural India, Mumbai, India
e-mail: psainath@ruralindiaonline.org

N. Bikkina and R. M. R. Turaga (eds.), *Climate Change Adaptation*,
https://doi.org/10.1007/978-981-97-1076-8_2

15

I want to discuss the idea of climate change through the telling of stories. Climate change is not on the public agenda. It is not in the top hundred topics of interest in this country. It is not in public discourse anywhere as much as it should be. We all know that it should be. But it simply is not. We can conduct any study we like of our local or national newspapers and find out how little attention is given to environment, let alone climate. The second thing I want to say is that climate change is not class-neutral. It affects some people more than it affects others. I want to read to you a quote from the UN Rapporteur on climate change. He had just stepped down two months ago from the post. He had the courage and the wisdom to state this while he was still the UN Special Rapporteur for poverty and human rights. So Philip Alston said this. "We risk a 'climate apartheid' scenario where the wealthy pay to escape overheating, hunger, and conflict while the rest of the world is left to suffer. The risk of community discontent, growing inequality and ever greater levels of deprivation among some groups likely stimulate nationalist, xenophobic, racist and other responses. Maintaining a balanced approach to civil and political rights will be extremely complex and difficult (*The Guardian*, 2019)." Prof. Prabhat Patnaik added to this saying "…well, apartheid was always the plan."

You know how you create smart cities and exclude 98% of the existing population of the city in creating little Walhallas and Xanadus. You can see this happening in Indonesia at a much faster scale than it is visible in most places. Jakarta sinks in the fast rising sea. Elite simply plan to move to new areas in the Borneo forest. That is going to devastate and destroy a very rare ecology, a very rare environment, a very fragile forest system, some of the most ancient tribes in Indonesia and the world and wipeout habitat of the famous orangutan and other species which are very special to that part of the world.

Climate change is not class-neutral, neither is it individual-neutral. To say that it is an equalizer, it affects everybody is as wrong as it is to suggest that COVID-19 has been an equalizer. We know in every society who were the worst affected. In India, we don't even keep data on this. But you can see it in the United States: the disproportionate number of deaths of black people, disproportionate deaths of people severely affected with co-morbidities, or direct deaths in COVID-19 (e.g., Reyes, 2020). Let us not lose the socio-economic context of what is unfolding because I maintain that the agrarian crisis and the climate crisis are deeply linked to our structural inequalities and to my mind, they are also very real moral crises. If we try looking at them with techno-fix approaches—that you correct this thing here, you create a better system there—we have caused 70% of the damage which I see in the country. So, the third point is that I want to avoid the techno-fix approach.

Fourth, I think the reason why it is so poorly positioned in public discourse is that the way it is covered in the media, the way it is told, it is like something far out there. There are two approaches. One is climate change is something that is happening in the wild fires of California and New South Wales in Australia. Climate change is something that is happening in the West Antarctic sheet. It is something happening out there in the Amazon forest. It is as if we are not seeing any effect of it. We are seeing it all around us. We are not taking cognizance of it. It is too abstract a problem for people to relate to. The other approach is that this is a seriously technical, highly complex, scientific matter to be dealt with by scientists, top level bureaucrats, and ministers for the Conference of the Parties (CoP) meetings in Paris or wherever. Again the ordinary reader or someone watching the television channel to the extent that this subject figures at all immediately moves away from it thinking this is too much for me, how do I relate to it? Where do I come into this? So we created a rather Brahmanical debate around it. Only a few wise people can speak and that excludes the public discourse.

The fifth point is that when we do this, we are missing out on a very real set of sources that can tell us a very great deal about what climate change is doing to us, what it will do to us, what it has done to us, how it did that process, and so on. These sources are the lived experiences of the ordinary people—of farmers, forest dwellers, fisher folk, all of them, their actual voices. In fact, if you look at it, you will find that the best research on climate change is those where the researchers have engaged with ordinary people; the best studies that I am aware of. Even in the IPCC there are many thousands of studies. You find that the best are those who have engaged with ancient communities. They have seen climate changes in different epochs, in different eras. So, we in the People's Archive of Rural India decided that we will try telling this through life stories of people and of particular regions.

Millets and Traditional Knowledge in Anantapur

I will start by showing you a couple of things that are happening with climate change from the Rayalaseema region, particularly from Anantapur because I guess over the last 25 years, I have spent more time there than in any other part of Andhra Pradesh. But I have taken one village called Dargah Honnur (Sainath, 2019, July 8). This is on the border of Karnataka. 60 years ago, this place was completely covered with millets. And then, the great idea came of bringing in groundnut, the cash crop. Contrary to common belief India was not a land of rice before the British divided it into wheat and rice. It is so simple. A very large set of populations, different cultures, and diversities consume millets of different kinds whether Ragi in Tamil Nadu or Fox Tail in Anantapur. East of India had more rice, but even there they had millets. In western India in Maharashtra it wasn't rice and wheat. It was Bajra and Jowar. These were the real cereals of the people. Now we moved everything toward mono-cultures or if you like duo-cultures—wheat and rice and the rest, cash crops. The thing that is outrageous in many ways is that even as we set out the process of introducing groundnut on a large scale, we were ignoring the devastation unfolding in the Sahel in Africa. We had the example of the Sahel before us. It turned into desert. By the way groundnut cannot gel well with shade. So you go about cutting off trees around it. You do all that.

Many people think of climate change as happening out there—either in the future or elsewhere. It is a physical, natural process. Swaminathan Anklesaria Iyer in Times of India argued that the best solution for environmental protection is to hand over the environment to the markets because when you want to make profits, you realize that you have to protect the environment. Well, it is stupid! But the fact is that we started this at random. Anantapur and Rayalaseema were famous for the Navadhanyalu—nine things you grew around the year. We removed the millets and started growing groundnut. Then came the bore well revolution. Not many people remember that in the 1980s, Government of India in all its wisdom launched a program called the Million Bore Wells Scheme. They actually exceeded the figure. So today when we live in an age where you have gray zone, dark zone, and black zone in different states, where you cannot open bore wells, where you cannot go deeper than this, it is easy to forget that we went out and put those bore wells there like it was a mission. It is also curious the way that affected Anantapur for instance.

Anantapur's carrying capacity is 70,000 bore wells. At the most conserva-tive estimate of how many bore wells there are now, it is 210,000—three times as much as we stated is the carrying capacity of the district. With the drying up of the land, the drying up of the soil, it is now going to be very difficult to reintroduce the millets, is going to be difficult to do a lot of things and it is also going to be very difficult to control your climate anymore. You have gone in for great deforestation; you have removed the surrounding fields. So you have more problems.

With the groundnut requiring the kind of water, the whole river which flowed through Anantapur has gone dead. Now what's happened is that Dargah Honnur and surrounding areas have become so much of a desert that Bollywood and Tollywood are shooting their desert fight scenes in Anantapur, no longer in Rajasthan at Jaisalmer and Barmer (Sainath, 2019, July 8). This is much cheaper. You can even stay in Bengaluru and go across and shoot there. It's raining sand in Rayalaseema in Dargah Honnur village. This land was covered in millets. Now there is some 2000 acres that is a desert. There was hardly a single bore over there 20 years ago says a farmer who has been cultivating in twelve and a half acres. It was all rain-fed agriculture. Now there are 400 bore wells in 1000 acres. There is one bore well to every three acres. It is pretty bizarre what happens to it thereafter.

By the way, the point about Rayalaseema is that you have destroyed the food security of the people of Rayalaseema. Tomorrow when there are great shortfalls, it is very difficult to reintroduce millets. It's a very complex process. It can be done only with lots of investment and with major state intervention. Unfortunately, even the knowledge of how to recreate the Navadhanyalu is lacking. The present generation does not have farmers who know how to do this. As in the next 25 years, the impact of climate on food production is going to deepen. In fact, Rayalaseema is going to find it very difficult to cope. That is Andhra Pradesh.

SUGARCANE AND WATER IN MAHARASHTRA

Now you take Maharashtra. There is this animal called the Gaur Buffalo. It is actually not a buffalo but an ox. It is known as the Gaur Buffalo. The British called it that. It is the largest bovine in the world. This is in the Kolhapur Wildlife Sanctuary (Jain, 2019, July 15). About 30 years ago, we started allowing mining around this area. This is the last great refuge of the Gaur Buffalo. We have Gaur here, in Assam and a couple of other spots. These two to three spots are the last refuge of the biggest bovine in the world—the male standing at seven feet, about 6.4 feet at the shoulder, a kind of magnificent creature. It is very important to that region in more ways than one, to the whole ecosystem. We started allowing mining over here, started allowing them to use the plateau overlooking the sanctuary in which the Gaur are a part. Now what has happened? The water sources are contaminated and have dried up. It is very strange that Kolhapur has been experiencing every year or the other year devastating floods alternating with total drought and water scarcity within the sanctuary areas because of the mining. Now these giant animals are out among the villages eating the food crop the people grow around them.

Meanwhile, Maharashtra's emphasis moved to sugarcane. Do you know that cultivating one acre of sugarcane takes 18 million liters of water? This is the Department of Agriculture's figure. Independent experts calculating the same data says it takes 21 million liters (Lee et al., 2020). Let's take the Department of Agriculture's figure. 18 million liters is equal to seven-and-a-half Olympic swimming pools. 21 million liters is equal to nine Olympic sized swimming pools. This is for one acre of sugarcane. In those same 18 million liters, we can grow 12 acres of Jowar, which is an important cereal in Maharashtra. You can grow 12 to 13 acres of Bajra. There is also a history of our addiction to sugarcane, to rice, and to wheat. The British came to your country. M.S. Swaminathan has explained this more than once. They did not recognize your very nutritious native cereals. They understood rice, they understood wheat. They emphasized the cultivation of those. They fed our cereals to their cattle and livestock. That is how you got the name cow-pea, chick-pea, and horse-gram. These are not accidental names.

For 50 or 60 years after that if you look at our economic surveys and calculations, we still call our own food species as crude cereals. There is nothing crude about Ragi. It is a highly nutritious millet. But they called them crude cereals and fed them to their cows, their birds, their poultry, and their horses. Then, of course comes the Sanskritization part where we think that the upper classes are all into rice and wheat. Rice becomes a status symbol. We should be eating rice. Actually rice was consumed in Rayalaseema on special occasions like festivals, happy days, celebration days, and wedding days. The staple was millet. In Maharashtra we have smashed the diversity by bringing in sugarcane. Incidentally two-thirds of Maharashtra's sugarcane is grown in drought prone areas (Lee et al., 2020). All that has combined to make life miserable for the magnificent

Gaur. Every day the fields of surrounding farmers are raided by animals that neither you nor the police nor anyone else can drive away. These are behemoths. When there are 200 of them around you, don't fool around with them. You accept your crop loss. Now the farmers in the surrounding area are not asking for these animals to be killed or anything. They don't blame the animal, by the way. It is very interesting. They speak to us; blame the forest department for having degraded the forest, having allowed the degradation of the forest which is driving these animals to their fields. You are having all kinds of human-animal conflict on the highways of Maharashtra around Kolhapur now because of this.

Nomads, Fisherfolks, and the Knowledge of Climate Change

India is home to the largest number of nomadic pastoralists groups. These are people who are at ease with the climate, who are masters of climate adaptation. From them we have an incredible amount to learn. But in fact, we are now busy destroying their climate. India, China, and Pakistan, all of them building border roads along that way. It is having a phenomenal impact on the Changpas. Do you know who the Changpas are? They are the people who breed your Cashmere goats and sheep. They are also the people who breed the Yak. They have made from their personal experience, huge studies of what happened. The Yak is an incredible creature. It is very comfortable in temperatures of 30 degrees minus. Minus 30 degrees Celsius, an animal that is okay with that. It can survive minus 40 degrees Celsius. When it comes to 12 degrees, 13 degrees plus the Yak is in trouble. New York Times carries an app which you can download (How much hotter is your hometown than when you were born, 2018). You can go anywhere. It is a 60-year app. You can go and see, for instance, what were the temperatures in Vizag 60 years ago? By the way we have put that link to the app in all our stories of climate change. You can find out how many hot days there were in a year in Vizag 60 years ago, how many hot days in a year there are in Vizag today and you will see the number of days, how much it has risen. There are places in India where the number of hot days above 32 degrees Celsius has risen by 80 days in a year. Changpas are telling you that their livestock are being degraded; they are telling you that the Cashmere wool is getting thinner because the animals have to adapt to warmer weather (Mukherjee, 2019, July 22). Now the Changpas (Fig. 1) are having to go higher and higher into the hills for grazing because the animals can only survive at those temperatures. What the heck they will find by ways of grazing material. These are the highest grazing grounds in the world at 19,000 feet. I think if we talk to our children, to our readers, to our public through stories, let the Changpas speak, let the Gaur Buffalos' neighbors speak, let the people of Dargah Honnur speak and if we listen, we are going to learn a hell of a lot more.

This story is from Wayanad. These are the coffee growers of Wayanad. A lot of coffee comes from those cool Kerala–Nilgiri bordering areas. I have spent a lot of time in Wayanad myself. This is done by our social media editor and people are thinking about climate. If you go and ask them what you think of climate change, they may not answer your question because they may not know that phrase. They know the applied effects of that phrase. Unlike many urbanites I find that farmers, agricultural laborers, honey collectors, and fishermen have a far more nuanced

Fig. 1 A Changpa with his pashmina goats (*Source* Mukherjee, R. (2017 February 8). The Changpas who make cashmere. People's Archive of Rural India. Retrieved on 20 September 2023 from the World Wide Web: https://ruralindi aonline.org/en/articles/the-changpas-who-make-cashmere/)

understanding, because they have to; by their nature they have to. They make a far greater distinction between temperature, weather and climate, whereas we in urban areas very often use these terms interchangeably. A variation in temperature, we very easily say weather is changing or climate is bad. And you find that in all the Indian languages, there are very clear terms for each of these—temperature, weather, and climate. They have an understanding of what each of those nuances mean. The coffee growers of Wayanad are already experiencing very severe production losses (George, 2019, August 5). They are talking to us about it.

Another group of people have started making their own solutions. One of the most fascinating experiments in this country is coming from ordinary people and scientists are supporting these experiments. The MS Swaminathan Research Foundation (MSSRF) is supporting a group of fishermen on the tip of the country in Ramnad. Now these fisher folk have discovered over the years many ways they have been affected by climate change. They are an encyclopedia of information on climate change without using the phrase climate change. That is our usage. It is an *angrezi* usage which they understand in different terms and we think that they don't know anything. They know hell of a lot more than you and me. And it is a question of how they are able to make use of that knowledge. First of all develop the humility to understand. There are people who know more about this than we do. And while the MSSRF is supporting these fishermen, what these fisher folk have done in Munnar and Pamban area is that they have created their own radio station (Muralidharan, 2019, August 12) to warn their people when out at sea, to give them signals, to tell them that our understanding is that the seas are going to be very rough for the next six to seven days, which by the way is affecting Lakshadweep very badly also. There is something very beautiful about this story—the woman Tamilian's humor. When our reporter was interviewing her, she was saying what fish were available 30 years ago and what fish are available today. She rattled off the names of 30 to 40 fish species. She knew which channel of the sea had got which fish. There are different channels in the seas, especially when there are three seas meeting over there. When you have the Indian Ocean, the Arabian Sea, and the Bay of Bengal coming into the picture, there are many channels. I love the way she put it. She said in this channel, you got these species, in that channel you got these species. She says for certain number of species you have to look for them in the Discovery Channel because they are not anywhere in these channels. Petrol and diesel have poisoned the waters, altered the taste of the fish. Her grandparents who used non-mechanized country boats had warned them that engine sounds chased the fish away; this petrol is going to poison the fish. She said that they were absolutely right. There is something to learn here from the traditional fishing communities. The amount of knowledge they are disseminating on their radio station is fantastic.

Marathwada was a heavily drought prone area. 105 years ago, cotton arrived in Marathwada. Prior to that it was like Rayalaseema—covered in millets. The day we shifted from millets to cash crop cotton, Marathwada's agro ecological balance was disrupted. Today it is sugarcane and cotton. And, one of the hottest parts of India where 43 degrees Celsius is very common in early summer, in April, is having hailstorms every week. The hailstorms are with hail so large that a significant number of cattle and sheep are dying. Hailstorms at 43 degrees Celsius wreck farming in Latur. There are old women farmers in Latur who will tell you what changes occurred, what we should not have done, and what we could do about it.

I showed you the Changpas in western Himalayas, the growers of Cashmere. An equivalent nomadic pastoral community in the eastern Himalayas is the Brokpas. By the way, if you see this picture (Fig. 2), this is not a dog. It is a baby Yak, who thinks that this herder is his mother. It follows him around. It's only seen this guy, since it was born.

The Brokpa are an extraordinary people. There is serious scientific research going on in terms of how they are trying to mitigate climate change. They have created a new hybrid of animal because now the Yaks have problems at warmer temperatures. Traditional cattle have problems at higher temperatures. They have cross bred the Yak and the Khot. The Khot are the highland cattle. They cross bred the Yak and the Khot and created an animal called the Zomo. They are also going higher and higher to newer grounds of grazing. They have revised their traditional migration paths because they are talking of what is happening with some glacier, with some snowfields, with some former greenery. They are adapting by changing their migration paths upward. These are people who migrate upward, not downward for their needs. There is a 13-year-old kid telling us when asking about good old days that happy days are now just nostalgia which of course, he picked up from hearing his old grand-dad or someone saying. But it is a very worthwhile saying.

And then we have the world's largest mangrove forest—Sundarbans. May be one-fifth or one-fourth of Sundarbans is in India, the rest is today in Bangladesh. All kinds of stupid ideas introduced into the Sundarbans

Fig. 2 A Brokpa with his baby Yak migrating to higher altitudes (*Source* Mukherjee, R. (2019 July 22). Perhaps we made the mountain god angry. People's Archive of Rural India. Retrieved September 2, 2023, from the World Wide Web: https://ruralindiaonline.org/en/articles/perhaps-we-made-the-mou ntain-god-angry/)

are devastating the largest mangrove forest in the world, and having its effect on food production and livelihoods. Again you will find people there who can explain it to you.

Even better, in the opposite climate we have the nomadic pastoralists, the Maldharis in Gujarat in Kutch. These are people who do 800 kilometer migrations. They are transhumance going at a particular time between particular areas. Have a look at them. They will tell you what is happening. They have a great understanding of what the prices of wool on the global market are doing to sheep in the Kutch.

THE UNEQUAL IMPACTS OF CLIMATE CHANGE

More dangerous is what is happening at the home base of rice. Hope you know that the first cultivated rice emerged simultaneously in the Yangtze Valley of China and in the Jaipur–Koraput region of Odisha. Today, can you imagine that in the last 15 years, this basket of hundreds and hundreds of rice species—even as late as the 1980s, 62 or 70 distinct species were there to find—is switching over to Bt Cotton, switching over to chemical intensive cotton, because the corporations are on the run from Maharashtra and Vidarbha where they have already done their damage. In Rayagada, Bt Cotton acreage has risen 5200 percent in 15 years. Rayagada–Jaipur region is one of the biodiversity hotspots on planet Earth. Your black rice, red rice, and brown rice, which are still cultivated and grown from 8000 years before, are disappearing at a fast rate. As you are bringing in cotton, there are immediately discernible, visible effects on the environment, the ecosystem, and the climate.

One of the most extraordinary communities in the world include poor women who are seaweed harvesters, who go out where the water is four, six, to eight meters deep. Seaweed makes millions of dollars for the big pharma industry. Women who risk their lives for it get a few rupees for each kilogram. This is worth thousands of rupees. Even your cooking jelly is made with this kind of seaweed. This is a most extraordinary essay of how the women are talking about their recognition of what is happening with climate—how the seas are getting rougher and how their work is getting tougher.

How many people know that there are seven to ten thousand farmers in Delhi? The city of Delhi has destroyed the Yamuna River. The Yamuna River is 1376 kilometers. Only 21 kilometers pass through Delhi but that stretch of 21 kilometers counts for 86 percent of pollution in the Yamuna wreaking disastrous effects on livelihoods, climate, on the river, on the killing of many tributaries to the river.

Churu in Rajasthan is a climate hotspot. In 2019, it was recorded and reported as the hottest place in the world at 51 degrees Celsius in the month of June (Joshi, 2020 June 2). But you know what? It is the coldest part of the plain areas in Rajasthan between December and January. People are at their wits end as to how to cope with these huge changes. They say that they can cope with the temperatures but what is happening is essentially that the summers are growing much longer. They don't mind the winter coming. They need the winter. But summers are now eight months.

In Thane the Adivasis, one of the oldest cultivators, are at a severe loss. In Lakshadweep the corals are in extreme danger. How many of us understand the importance of insects? In India, 40% of our identified pollinator species are in steep and sharp decline. That is going to have a very severe impact on your agriculture, your horticulture, your food security, and everything else.

Conclusions

The reason I took you through these is that rather than tell you that this is the IPCC's agenda for reform or this is what governments are unable to deal with, I want you to look at what is next to you. You have Odisha on one hand; you have Anantapur on the other. The idea is that we start learning about what is happening to our planet, what is happening to our immediate vicinity. My hope is that the first thing we learn is how serious the role of human agency is, in this entire issue, how serious is our own complicity with building the kind of social and economic systems, consumption systems that exacerbate the kind of changes we have seen, which in turn impact so seriously on the livelihoods of people. My hope is that through these stories we get to know the knowledge of ordinary people. Many of these are ancient communities. Even those which are not are suffering in various ways. The idea is that we learn from the everyday lives of everyday people. Second, that climate change can be made a comprehensible debate where every citizen feels involved. May be if we could do that, we could certainly bring climate change as an issue higher up on the agenda of public discourse and make it a word in everyday use. Please go and look at how many political parties have climate change as an issue on their manifesto. How many times the word climate change is being uttered in the Lok Sabha or the Rajya Sabha? May be by going back to people, by communicating to people, amplifying those voices from the ground, we can have some impact on that.

References

George, V. (2019, August 5). Why is the climate changing like this? *People's Archive of Rural India*. Retrieved September 2, 2023, from the World

Wide Web: https://ruralindiaonline.org/en/articles/why-is-the-climate-cha nging-like-this/

How much hotter is your hometown than when you were born? (2018). *The New York Times*. Retrieved September 2, 2023, from the World Wide Web: https://www.nytimes.com/interactive/2018/08/30/climate/ how-much-hotter-is-your-hometown.html

Jain, S. (2019, July 15). Buffaloed by the Climate in Kolhapur. *People's Archive of Rural India*. Retrieved September 2, 2023, from the World Wide Web: https://ruralindiaonline.org/en/articles/buffaloed-by-the-climate-in-kolhapur/

Joshi, S. (2020, June 2). Churu: Blowing hot, blowing cold—Mainly hot. *People's Archive of Rural India*. Retrieved September 2, 2023, from the World Wide Web: https://ruralindiaonline.org/en/articles/churu-blowing-hot-blo wing-cold---mainly-hot/

Lee, J. Y., Naylor, R. L., Figueroa, A. J., & Gorelick, S. M. (2020). Water-Food-Energy challenges in India: Political economy of the sugar industry. *Environmental Research Letters, 15*(8), 084020.

Mukherjee, R. (2019, July 22). Perhaps we made the mountain god angry. *People's Archive of Rural India*. Retrieved September 2, 2023, from the World Wide Web: https://ruralindiaonline.org/en/articles/perhaps-we-made-the-mountain-god-angry/

Muralidharan, K. (2019, August 12). Today we seek those fish in Discovery Channel. People's Archive of Rural India. Retrieved September 2, 2023, from the World Wide Web: https://ruralindiaonline.org/en/articles/today-we-seek-those-fish-in-discovery-channel/

Reyes, V. M. (2020). The disproportional impact of COVID-19 on African Americans. *Health and Human Rights Journal, 22*(2), 299–307.

Sainath, P. (2019, July 8). It's raining sand in Rayalaseema. *People's Archive of Rural India*. Retrieved September 2, 2023, from the World Wide Web: https://ruralindiaonline.org/en/articles/its-raining-sand-in-rayalaseema/

The Guardian. (2019). 'Climate apartheid': UN expert says human rights may not survive. Retrieved September 2, 2023, from the World Wide Web: https://www.theguardian.com/environment/2019/jun/25/climate-aparth eid-united-nations-expert-says-human-rights-may-not-survive-crisis

Ethical Adaptation to Climate Change and People-Centered Development

Abstract Climate change is fundamentally an ethical problem that has moral connotations. In this context, adaptation strategies need to be people-centered, where the ensuing processes and outcomes are capable of ensuring equality and justice, care and empowerment for those who are disadvantaged and vulnerable in society. Such an approach therefore involves not only strengthening the adaptive capacities of vulnerable groups, but also deconstructing certain dominant narratives surrounding climate change and climate change adaptation. Prevalent metanarratives on climate change do not emphasize adequately on the everyday life struggles and knowledge systems of poor and marginalized actors. Moreover, present day planned adaptation strategies tend to ignore the cultural, social, economic and political contexts shaping people's marginalization and vulnerability. They tend to discount the underlying social inequalities and inequities that shape these diversities. Instead they have the potential to reinforce existing vulnerabilities and livelihood insecurities. There is a need for knowledge, actions and movements situated from the standpoints of the marginalized actors. We need to rely on the strengths of people, their cultural narratives, local knowledge systems and

S. D. Santha (✉)
Tata Institute of Social Sciences, Mumbai, India
e-mail: sunilds@tiss.edu

© The Author(s), under exclusive license to Springer Nature Singapore Pte Ltd. 2024
N. Bikkina and R. M. R. Turaga (eds.), *Climate Change Adaptation*,
https://doi.org/10.1007/978-981-97-1076-8_3

29

wider social networks. For many poor and marginalized communities, the only power they have is their local knowledge and their solidarity networks. Nevertheless, local knowledge systems are both situated and dynamic. The historicity of marginalization and resource inequities among those vulnerable to climate change can never be ignored. Further, we have to give attention to the politics of difference in the production of knowledge, where social interactions are usually routinized across intersectional dimensions of class, gender, race, caste, ethnicity, region, religion, language or occupation and their natural environment. The arguments presented here would be substantiated through the life stories of small-scale fishing communities in India.

Keywords Fishing communities · Climate change · People-centered development · Knowledge systems · Distributive justice

As development scholars and practitioners, we need to affirmatively prop- agate certain key values that would facilitate the designing of ethical adaptation strategies to climate change. In this regard, we need to critically examine climate change adaptation from diverse knowledge perspectives including the circulation and politics of knowledge. Are we merely guided by certain static universal knowledge frames by pointing out to the global discourses out there or are we capable of transforming the discourses at the grassroots beyond the local? Or are there any inter- faces happening between the global and local constructions of knowledge? And if so what is the nature of such knowledge interfaces? Examining the writings of several thinkers like Appadurai (1990), I could anticipate the inherent politics and dynamics of contesting knowledge systems in specific fields of practice. However, my argument here is that such understanding and debates are extremely limited in the field of climate change adapta- tion. Climate change discourses in the Global South are largely driven by certain metanarratives (Chaturvedi, 2014). My interest was largely concerned towards the implications of these metanarratives on the local livelihood practices of people and their everyday constructions of knowl- edge. However, such an inquiry further led to my conviction that we need to recognize the situated nature of knowledge or situated knowledge.

Justice, Care and Solidarity as Three Pillars of Adaptation Strategies

Three important values of justice, care and solidarity remain the theoretical base of my above analysis and reflection (Santha, 2020). As development practitioners, when we go to the field and start working with communities as well as with the state, I believe that the values of justice, care and solidarity turn out to be very crucial in determining the success of adaptation projects. When we talk about justice, care or solidarity, we are actually talking about an ethical adaptation to climate change. When you talk about justice and equity we are not talking about distributive justice alone. We have to involve more with procedural and structural justice dimensions as well. When we talk about relationships of care, we are valuing trust and reciprocity in human–non-human relations and a kind of sacredness that exists in dealing with nature. Caring solidarity ensures that such relationships and networks are nurtured, decentralized decision making and ownership of resources and management of commons, etc., are preserved. When you look for an ethical model like this, there are also chances that we would be thinking of the intersectional, intergenerational and interspecies needs. As an academic and action researcher, I see this lens as very important for climate change adaptation.

Talking about climate change today, it is also important to realize that any form of adaptation has to recognize the dynamic interconnectedness of social-ecological systems. So, distributive justice alone may not work here. Universal solutions may not work here because adaptation decisions are context specific. At the same time when we talk about people centeredness in adaptation, the problem is that we need to look at the intersectionalities within the so called humans—be it gender, be it caste, be it class and all those kinds of variations and diversities, including intergenerational. There is also the relationship between humans and non-humans that is bio-diversity specific. Simultaneously we need to understand the situatedness of knowledge amidst its circulation. Feminist frameworks on the situatedness and the circulation of knowledge thus add meaning to the above assumptions (Code, 2006; Haraway, 1988; Santha, 2020). In this regard, I carry forward the assumption that any kind of knowledge, be it local or global, is plural, dynamic and nonlinear. The relevant question then is how are these perspectives going to add value to the debates on ethical adaptation to climate change.

KNOWLEDGE, ETHICS AND ADAPTATION

The root causes of climate change also signify the existence of an ethical problem that has strong moral implications too. Both global and local discourses surrounding climate change showcase problems of inequity, or injustice or historical oppression and exclusion. These are often the reflections and continuation of those debates that critically examine problems associated with present day development models. So when we analyze the modern development discourses, there are certain unethical values that are attached to the word 'development' and its constituent processes. There thus exists a metanarrative of climate change (that makes us realize that climate change is not restricted to a single scientific or an epistemic community but had crossed those borders a long time back) shaping adaptation practices in India, nations in Africa and Asia. These practices are shaped by western scientific knowledge, especially if you look at adaptation. Mitigation had long back itself become consolidated by economics as a key discipline of shaping innovation. But in adaptation it is largely community centric or largely region centric (Chaturvedi, 2014).

You can see that still the science politics are driven by experts and institutions in which these experts have been located and it has its own power structures and the way knowledge is circulated and legitimized has its roots in western scientific knowledge. Now that kind of scientific knowledge may not suit the context in which local communities exist and how they interact with the resources and how their everyday livelihood practices are shaped. Chaturvedi and Doyle (2015) state that in its very construction as well as existence, climate change is rooted in the Global North and therefore you can see that global adaptation models talk of transformative adaptation models. What are these transformative adaptation models or other kinds of global adaptation models that are brought out in the name of the best practices? These all have a kind of a dominating knowledge system that always may not match the local or even specific global contexts. Further, these emerging discursive practices also ignore the environmental justice perspectives of the South and then the language is a mix of both. So the actors in the South almost can get carried away with some of these articulations and practices. We in India had once witnessed such pitfalls with universal implementation of the sustainable livelihood framework as a developmental model, and also as a panacea to all poverty related issues. It is in such larger politics of the global that climate change adaptation encounters the local.

When we go forward, we need to certainly deconstruct the metanarratives because they don't look into the everyday life's struggle of poor, marginalized, rural communities who are dependent on natural resources for their survival. Their livelihood practices and knowledge systems are often excluded or not at all recognized. At the same time we forget the historicity of how people have become vulnerable, how resources have become degraded and how neglected are the micro and macro contextual factors. If you look at Wisner et al. (2004) political economy model, it clearly says it is lack of resources, decision making structures and power shaped by both political and economic ideologies that are the root causes of peoples' vulnerability. Present day adaptation frameworks somehow fail to recognize these; while these historicities are also present in the narratives depicting people's local and indigenous knowledge systems. This is not only about the knowledge of flora and fauna. There is a mistaken understanding that local knowledge systems or indigenous knowledge systems are how people relate to the flora and fauna. No! It is about the historicity of oppression, people's agency and transitions in adaptive capacities over a period of time; as to how people are surviving in much weakened environments and how they have resisted different forces to sustain their resources and livelihoods. All these are part of local knowledge systems.

Fishers' Narratives

There is a need for deconstructing the metanarratives here. Today, we don't listen to or recognize any voices among the people who are actively affected by climate change though there is a lot written and said about it. From a practice point of view, the very same people become the victims of climate change or you can call them climate refugees and many times there are silent witnesses to exclusionary, oppressive and regressive interventions. How do we address these inequalities? That is where we need to look at the local knowledge system as an alternative, where the knowledge, action and livelihood practices are recognized from the standpoint of the marginalized actors. To illustrate this further, let me share some field transcripts here. This is based on a study done with traditional fish workers in Kerala. This narrative indicates how they remember as to how hazards were earlier.

If we say that the surge will happen on this date, it may get delayed, but will certainly happen. Sea surge used to be a regular phenomenon, which happened every year…during June-July. The surge will remain for 26-30 days. We predict the hazards by looking at the changes in nature, wind, and sea. If the waves reaching the shore pull the sand heavily from the shore towards the west, it indicates the beginning of the surge. This is locally called 'arappan'. It can erode the whole land and destroy our homes. We have lost several huts during these storm seasons… . In the month of Edavam, the sea will be usually calm. But, on the 28th day with the rising of the star, the sea surge will occur. The sea will then become turbulent with full of foams and waves.

It is not as if sea surge is new to the community. They are already exposed to it. But they had predictability to this. They say that:

In the month of Medam, the sea will come from both the south and the north. Simultaneously, there will be lightning striking at the west. The surge will occur from that spot of lightning. Then the sea will turn rough in the west. The waves will be striking intermittently forming bubbles and waves. The sea snakes during such times will be floating at the surface of the sea cuddling each other like a ball. These are all strong indications of an upcoming surge.

So the nature of the earth, the direction of the waves, the location of the lightning, the behavior of sea snakes and the like include several other indicators of local knowledge that serve as an early warning system. But what is happening to the utility of such a knowledge system in the context of climate change? An elderly fisherman states:

Earlier we were able to forecast the changes in nature in advance and more effectively. Those days, the normal seasons could be categorized as three months of winter, three months of summer, three months of heavy rains, and three months of moderate and intermittent rains (idamazha). We could understand this pattern and forecast the changes that were about to happen. These days it is not possible to forecast. The changes happening to nature are within seconds and happen immediately… . These days fear lingers that anything can happen at any time. The probability of incurring losses is also very high.

Another group of fish workers say that:

These days no one can accurately define what 'risk' is. Whatever happens unexpectedly is a risk. Today, the sea surge or coastal flooding occurs without any indication and these are risks. They can also manifest in terms of unanticipated wind and rains, accompanied by extremely dangerous surges. The climate has changed. But the changes in nature are also due to human intervention. Burning waste and an increase in the usage of plastic have resulted in changes in the atmosphere. The impact of these changes is mostly felt in the sea. Thus, the atmosphere gets heated up and changes in nature occur. The beginning of any risk or extreme hazard event takes place at the sea. Whatever we do, its impact is felt in the sea. Even when forests are destroyed, its effects are felt in the sea. The whole water cycle leading to rains gets affected. Even the action of testing the purity of gold affects the water cycle.

They relate their situation similar to that of the butterfly effect in science that all actions finally impacts the sea. Slowly the narratives change from their tradition to the uncertainties, to how modern forces, the formal forces like the state agencies are largely shaped by the scientific knowledge and interventions of the west:

Earlier, such hazard events were less. We are not saying that it was never there. It was minimal, and people of the olden days knew in advance of an impending kolaru. The weather of the olden days was also different... . Those days the risk was different. It was death due to starvation and poverty. With the modernization of fisheries, we are able to overcome starvation. Today, the risk of extreme hazard events has increased. These days we use fiber boats fitted with engines and venture into the sea, even if there are chances of hazards. Earlier it was the catamaran, and therefore we were reluctant to go if the sea was rough. Today, everything is a new world, and unpredictable. And it seems as if we also know nothing.

Here they criticize how the state or the other organizations intervene in terms of adaptation like constructing a sea wall. Every year they have to maintain the seawall because it collapses and this is where they see the value of their traditional practices:

Before the construction of this sea wall, we used to dig diversion channels and fill gunny bags with sand to prevent coastal flooding. It was a whole community effort every year before the monsoons. These practices have stopped after the sea wall was built. The sea wall that has been constructed here is highly unscientific. The height and width of the structure needs to be relooked. As surges have increased, our shore has also eroded. The sea walls prevent us from resting the boats in our shore. Therefore, we rest them in the harbor. To reach the harbor, we have to take a bus. The expenses for all these are very high.

They use the legend or myth of Bhima to compare the present sea wall construction and other concretized or structural adaptation strategies:

Mother Sea had a brother called Bhiman. He once invited his sister for his daughter's wedding. However, the sea said that she would not attend the wedding, as she was apprehensive whether Bhiman and others attending the wedding would be able to bear his arrival. However, after continuous insistence from Bhiman, the sea decided to attend the wedding. In the meantime, Bhiman constructed very strong palaces and other physical structures to withstand the sea's arrival. However, all these structures collapsed as the sea entered the palace. Everyone died and everything in the land was destroyed. There was no wedding, there was no Bhiman!

So there are also myths and legends that are acting as a critique of this formal knowledge system and formal interventions. We need to lend an ear to these myths, legends and local knowledge to actively design context specific adaptation strategies.

Conclusion

We need to respect local institutional arrangements and resource management practices. We cannot take away the resources, which are the basic livelihood resources of the local communities in the name of adaptation. Then it results in their pushing beyond margins. So local culture, knowledge systems and livelihood practices need to be recognized where their voices are heard and practiced.

But I am also cautioning against going towards the romanticization of local knowledge, because at this very moment, local knowledge itself is not a homogenous entity. It is highly context specific; it is situated. It is context specific to temporal and spatial dimensions. It is intersectional. Women have specific knowledge; across different occupational and caste and class segments you can find different knowledge pertaining to a particular function or an activity or a resource system itself. All this also accounts for the politics of difference. There is a hidden power structure. Everybody is not equal even in a local community. There is patriarchy. There are other forms of caste inequalities and class inequalities. So we need to address even those kinds of exclusions and inequalities when we are talking about knowledge and adaptation. Unless we recognize those partial, plural and intersectional contexts, adaptation can end up just like a formal adaptation strategy itself. We need to recognize the politics of difference and hidden power structures; but at the same time try to work out on values of justice, care and solidarity. Action research and reflective practice are important means of using local knowledge and combining it with scientific knowledge to bring about incremental changes.

REFERENCES

Appadurai, A. (1990). Technology and the reproduction of values in rural western India. In F. A. Marglin & S. A. Marglin (Eds.), *Dominating knowledge: Development, culture and resistance* (pp. 185–216). Clarendon Press.

Chaturvedi, S. (2014). Marginalized coastal communities and struggles for social innovation in India and China: Towards a micro-geopolitics of environmental sustainability in the era of climate change, Working Paper Series, India China Institute. https://indiachinainstitute.org/wp-content/uploads/2014/04/Chaturvedi_Marginalized-Coastal-Communities-and-Struggles-for-Social-Innovation-in-India-and-China_WPS_2014.pdf

Chaturvedi, S., & Doyle, T. (2015). 'Climate borders' in the Anthropocene: Securitizing displacements, migration and refugees. In S. Croft (Ed). *Climate terror: A critical geopolitics of climate change* (pp. 109–131). Palgrave Macmillan.

Code, L. (2006). *Ecological thinking: The politics of epistemic location* (pp. 40–41). Oxford University Press.

Haraway, D. (1988). Situated knowledges: The science question in feminism and the privilege of partial perspective. *Feminist Studies, 14*(3), 575–599.

Santha, S. D. (2020). *Climate change and adaptive innovation: A model for social work practice.* Routledge.

Wisner, B., Piers, B., Cannon, T., & Davis, I. (2004). *At risk: Natural hazards, people's vulnerability and disasters.* Routledge.

Farmer Knowledge and the Politics of Climate Resilience

Marcus Taylor

Abstract It is frequently argued that local or traditional farmer knowledge must play a vital role in shaping climate-resilient agricultural development in southern India. At present, however, the Indian agricultural extension system has an uncomfortable relationship with farmer knowledge. Typically, it has pursued a relatively top-down and linear technology transfer approach in which new technologies and cultivation methods are designed off-farm and promoted as innovations to displace local practices. Recently, this mode of extension has been challenged by approaches that foreground local knowledge, networks, and experience, including the validation of landraces and cultivation practices that minimize external inputs. Drawing on material from three different case studies of agricultural extension projects in Telangana and Karnataka, we address the power relations and livelihood strategies that underpin the interaction between these different bodies of knowledge in rural south India. To explain what is at stake, we emphasize how the category of local farmer knowledge is often presented too narrowly. Local farmer knowledge is not solely concerned with cropping choices, landraces, cultivation practices, etc. It

M. Taylor (✉)
Department of Global Development Studies, Queen's University, Kingston, ON, Canada
e-mail: taylorm@queensu.ca

© The Author(s), under exclusive license to Springer Nature Singapore Pte Ltd. 2024
N. Bikkina and R. M. R. Turaga (eds.), *Climate Change Adaptation*, https://doi.org/10.1007/978-981-97-1076-8_4

is simultaneously knowledge of how to manage the accompanying social relations that underpin agricultural livelihoods including complex market transactions with multiple agents, credit/debit relations, labor management, and networks for social learning. Acknowledging this expanded framework of farmer knowledge is vital for any meaningful pursuit of rural resilience.

Keywords Resilience · Farmer knowledge · Livelihood · Knowledge systems · Agricultural extension

This collection of papers pivots upon the idea of traditional wisdom as a generative concept to start our discussions. It is also one that requires unpacking given the plethora of debates around the nature and roles of traditional, local, and/or indigenous knowledge (Gomez-Baggethun, 2021). Definitions of these forms of knowledge typically build from an emphasis on situated knowledge—the idea that these forms of knowledge are place-based and culturally bound and that they are primarily generated through experiential learning, that is learning situated in the material, social, and cultural contexts of particular places in particular times. This emphasis on local knowledge is typically contrasted with what is held to be modern or scientific knowledge (Agrawal, 1995). The latter is often held to be abstract, objective in the methodologies it uses, and universal in its application. In short, scientific knowledge is presented as not being bound to time and space in the way that local knowledge is.

What we presently see is that the climate change adaptation discourses, seek—in theory—to combine these two forms of knowledge to try to get the best of both worlds (Mustonen et al., 2022). On the one hand, climate change adaptation is envisaged to be driven by modern scientific knowledge that sets the broad parameters for conceptualizing climatic threats and impacts that must be adapted to. However, in its journey into particular cultures and spaces of adaptation, scientific best practices are argued to require partial synthesis with locally situated knowledge. Ultimately, through this marriage of forms of knowledge, there should be a kind of embracing and enrichment of both together to make the former more applicable and more effective. We see these kinds of arguments in many international discourses coming from the IPCC, the UN, the World Bank, etc. We also see this in the national discourse in India such as, the National Innovations in Climate Resilient Agriculture (NICRA) program run by the Indian Council of Agricultural Research (ICAR). Put simply, this idea of the complementarity of these two forms of knowledge is now a staple assumption in the climate change adaptation discourse: one complements the other to make both more efficient.

The picture, in reality, is often a lot more complicated than this idea of a seamless integration between two distinct yet enriching forms of knowledge. While we might aspire towards a frictionless meeting of the two, in practice we find that the ways that these two forms of knowledge meet are often bound up in hierarchies; they are attached to certain forms of power, both representational power about who has authority to define the questions of climate change adaptation, vulnerability, resilience and how we address them; and also the material side of adaptation where climate change adaptation comes with money and other resources. Put simply, the mobilization of different forms of knowledge to precipitate community action typically comes armed with correlated forms of financing, subsidies, inputs, and institutional leverage.

SCIENTIFIC VS. LOCAL KNOWLEDGE AS BINARIES

In approaching these questions critically, one of the first things to note is that the divide between modern, universal scientific knowledge and the idea of locally bound, situated knowledge does an injustice to both sides (Agrawal, 1995). On the one hand scientific knowledge is not quite so abstract and universal. Rather, as the Science and Technology Studies (STS) literature has convincingly shown, scientific knowledge is based on particular methods of knowledge production that themselves are place-based and are situated (Goldman et al., 2011). However, despite being produced in particular places and specific political contexts, scientific knowledge typically obscures the conditions under which it is produced to project its universal nature. As a result, in the field of agronomy for example, we see a lot of scientific knowledge that isn't quite so universal once it goes out and meets the complexities of the real world in all its diversity (Sumberg & Thompson, 2012). This is why we see so many new agricultural technologies that perform well in the test fields of the agricultural university and other controlled methods of ascertaining what works best in objective terms. However, these scientifically proven innovations turn out to be rather different when they meet the fields of real farmers operating under distinct environmental, social, and cultural conditions. All the certainties about how good an innovation is can melt away pretty quickly. In my own research on agricultural innovations in South India, I have witnessed the complexity of these knowledge politics all too frequently (Taylor et al., 2020).

And on the other hand, the idea of local knowledge is also prone to simplifications and romanticizations. On the one hand, the idea of local knowledge sometimes isn't quite so local or culturally bound as portrayed. Local, situated knowledge is always produced in relation to different knowledge bodies that are extra-local and therefore it is often drawn into an active relationship with scientific knowledge. As a result, local knowledge is not stable but rather is a constantly changing, dynamic form of knowledge. For example, in Mandya district, I was referred by local farmers to an important local rice landrace dating back centuries to time immemorial that served as a cultural market of local knowledge and practice. On investigation, it turned out that this 'traditional variety' was an early Green Revolution variety introduced in the mid-1960s. The product of external, scientific knowledge had been transformed into an artifact of local knowledge. As such, local knowledge has never been just about the local. This is not to say that we can't differentiate between situated knowledge and scientific knowledge. However, we need to be more than a little careful with these categories rather than positioning them as fundamentally dichotomous.

Varieties of Traditional Knowledge in Agriculture: Three Case Studies

On this basis, I want to talk about the knowledge politics of climate change adaptation, particularly the messiness of how these knowledge dynamics play out on the ground in the context of public agricultural development strategies. Within this field we often see sharp struggles over whose knowledge counts and whose knowledge can be mobilized through attachment to financing, money flows, and resources that rural development and climate change adaptation projects bring (Mosse, 2014). This site of struggle has been raised repeatedly in my own research across three studies of agricultural innovations bought into communities from the outside by public authorities. In conjunction with my colleague Suhas Bhasme from the Tata Institute of Social Sciences (TISS), we have looked at the system of rice transplantation in the district of Mahabub-nagar, Telengana (Taylor & Bhasme, 2019). Then, in Karnataka, we looked at hybrid rice varieties in Mandya district (Taylor et al., 2020). Finally, in Karnataka again, we examined climate-resilience projects in Tumkur district under the rubric of the NICRA program that sought to create model villages that displayed state-of-the-art climate adaptations (Taylor & Bhasme, 2021). I will draw on some examples from these cases as each of these projects was founded on the idea of public technology transfer of innovations into spaces where they weren't present before.

Mandya is known as being the 'sugarcane capital of India'. That said, rice cultivation is widely present across the lands irrigated by the Cauvery River. Our research here examined a project run by a network of scientists, researchers, and agricultural extension officers who sought to generalize a new hybrid rice variety KRH-4. The latter had been created to unify the higher productivity of hybrid rice with a locally adapted variety that could prosper in local agro-ecosystems. For those advancing the project, this new hybrid was intended to be a dynamic force of change in the agricultural environment. This is a very typical agricultural moderniza-tion strategy: agronomic scientists create the variety through advanced breeding techniques and distribute it to specifically chosen villages where they hire local model farmers to demonstrate it. This is typically accompa-nied by considerable fanfare and press attention while subsidies are given to farmers to encourage adoption. Ultimately, the idea is that the new variety will be proven to work in this locale and then diffused outwards by word of mouth and farmer example.

Importantly, this attempt to diffuse a new hybrid variety was occur-ring in a region that has seen a wide scale deskilling in the sense that agriculture had shifted away from the kind of situated knowledge and practice that previously existed towards a situation where farmers were increasingly reliant on outside inputs and outside expertise from agricul-tural extension officers, input dealers, etc., to perform many of the tasks of farming (Stone, 2007). There is of course a shift in inputs inherent to this knowledge transition. Not only does the generalization of commer-cialized hybrid seed require repurchasing on a yearly basis, but it also simultaneously requires more advanced usage of synthetic fertilizers and other inputs. As a result, farmers are becoming increasingly dependent on storekeepers for knowledge about which pesticides and fertilizers they should use. They literally take cuts of a plant that demonstrates signs of disease to input stores and ask storekeepers as to what product they should use—a role storekeepers are only too happy to play! This deterioration of effective local knowledge represents a deskilling process where farmers become more dependent on knowledge produced outside of the farm and the community. As a consequence of this deskilling, new power relations formed between extension networks, store owners who become really important in these areas, and also certain model farmers—often socioeconomically more affluent—who also give advice to farmers. Many of these model farmers are closely linked to public agricultural extension networks and benefit from subsidies through them. Like storekeepers, they can become greatly empowered through this process.

Such deskilling, however, provoked a counter-reaction from other farmers that sought to mobilize a different form of situated knowledge that is much more akin to traditional knowledge. Specifically, there were three particular model farmers in the area we looked at, each of whom proclaimed to be working with traditional knowledge in different ways. Each was tiered into different social networks at the regional level that shaped the kind of knowledge production that they were involved in and how they distributed that knowledge. First, one such farmer represented his rejection of outside knowledge and techniques as a return to traditional wisdom. This farmer, drawing heavily on Subhash Palekar's idea of natural farming emphasized agriculture as a spiritual endeavor, in which the use of technology was seen as something of a moral failing. Pointedly, this farmer argued that neighboring farmers who were using tractors and input-intensive cultivation strategies were lazy, alienated from the truth of farming, and had lost all substantive connection to the soil. This farmer demonstrated an impressive number of different varieties of rice that he himself was preserving. His networks that facilitated his knowledge production were broadly regional in which networks of farmers promoted the Palekar model of agriculture. In so doing, this farmer fostered a kind of knowledge that clearly was falling on deaf ears among his immediate neighbors. The failure to catalyze a more collective approach was unfortunate because following a more agro-biodiverse form of farming when surrounded by fields that are using chemicals limits the kind of agroecological synergies that you were planning for your crop.

The second kind of model farmer embraced a different kind of situated local knowledge. We might call him a modern-traditional farmer. In turning away from a highly chemicalized, scientific model of farming, this farmer embraced agrobiodiverse farming models predicated upon local varieties, seed saving, and sharing. While emphasizing working with local knowledge, this farmer was tied closely to networks of knowledge exchange that were both national and transnational in scope. He was very aware of international discourses around agroecology, traditional farmer knowledge, and seeds through national and international networks that were broadly aligned with Via Campesina. This was in contrast to the first farmer who was representing his actions as an embrace of traditional wisdom through a discourse of returning to an authentic form of *desi* farming. In many respects, both farmers were following similar approaches, but the knowledge politics around each were distinct.

The third model farmer that we found to be notable in terms of knowledge production was closely tied to agricultural extension networks, who had been a model farmer for the local Zonal Agricultural Station demonstrating the latest innovations for which he received various awards. Interestingly, however, when we interviewed this farmer some two years after the extension agents used him to demonstrate their new hybrid rice variety, he turned his back on the formal scientific knowledge networks that were trying to generalize input-intensive modernized agriculture. Despite at one point being a role model for modern agriculture, he had mobilized his local network of farmers around local varieties of rice and sugarcane, which he described as a return to the way that his parents farmed. But any sort of direct intergenerational transfer of knowledge could only have been partial. Instead, this farmer and his group were busy creating a new form of local, situated knowledge appropriate to their agroecological and social circumstances. This is a farmer who also had worked with agricultural extension officers closely, took bits and pieces of what he viewed as positive about new sowing techniques, weed management, pest prevention, and so forth, and amalgamated it into a hybrid form of knowledge and practice. Importantly, this was done to create a form of agriculture that is extremely risk-averse. This farmer turned his back on cutting-edge agricultural technologies advanced by the local university research system because they constituted too great a risk in conditions of high indebtedness, unreliable rains, and strained ground and surface water provision.

Power Relations and Local Knowledge

What became clear through all these examples of contrasting knowledge politics is that we need to position knowledge production within its wider political-economic context. Specifically, for the farmers involved in all three examples above, the key motivation for embracing this alternative form of knowledge and practice was to try to escape from relationships of debt and dependency—both within and outside the community—that formed the bedrock of commercial agriculture in contemporary rural India. In particular, farmers sought to develop a localized form of agricultural practice predicated on alternative agricultural techniques that, while having less overall yield potential, avoided the need to take on debts from various rural social agents such as moneylenders, input dealers, landowners, and microfinance or bank loans. The production of local knowledge was therefore not specifically knowledge of agroecological conditions and techniques. Rather, local knowledge was equally produced with respect to the social relationships that, on the one hand, make agriculture possible yet, on the other, can be exploitative and embed social hierarchies across the rural sphere.

This is important for how we think about climate change adaptation. Often we encounter this idea of climate change adaptation as driven by technical changes to agriculture to mitigate problems experienced by farmers in their fields about what happens to crops. This, however, is not adequate to capture the social dynamics of adaptation. For many farmers, climate change expresses itself most sharply in the social relationships of credit and debt: to whom, at what cost, with reference to both the social obligations that taking on debts can entail alongside the interest rates paid. It is the real lever of power and it is something to which many farmers orientate their agriculture. It involves the generation of situated knowledge but in the context of trying to avoid situations that threaten them with unsupportable debt and relationships with middlemen, commercial agents, moneylenders, and landlords. Thus local knowledge is not simply about environmental knowledge in terms of good agricultural practice. It is also knowledge about power relationships through which agriculture as a practice is made possible. Creating new forms of knowledge is therefore about enabling both networks and practices that can shift these power relationships.

In returning to the foundational point of this volume, one important point about 'traditional wisdom' is that the idea of 'tradition' makes it seem to be fixed in time whereas local and situated knowledge is very dynamic precisely because it responds to these relationships highlighted across this paper. These knowledge systems are not just about agro-environmental knowledge or at least not in the narrow sense of knowledge pertaining solely to what goes on in people's fields. Rather, it is also about the social environment, about acting within a socio-cultural environment, which has particular hierarchies, dependencies, and power relations. For example, when discussing rural knowledge politics we need to be attuned to the gendered dynamics of these relations. All the knowledge networks that I saw amidst both the modernist scientific approach but also the traditional agricultural knowledge networks were overwhelmingly masculine in the way that the primary actors, indeed sometimes the only actors, were male. We see this particularly on the scientific side wherein the agricultural extension officers were almost exclusively male; they worked with male model farmers to give knowledge to the male farmers on the ground. However, this was also present within some of the alternative agriculture knowledge networks as well. So these hierarchies are intersectional, forcing us to consider how producing alternative knowledge can positively reshape class and gender roles within rural environments.

References

Agrawal, A. (1995). Dismantling the divide between indigenous and scientific knowledge. *Development and Change, 26*(3), 413–439.

Goldman, M., Nadasdy, P., & Turner, M. (Eds.). (2011). *Knowing nature, transforming ecologies*. University of Chicago Press.

Gomez-Baggethun, E. (2021). Is there a future for indigenous and local knowledge? *Journal of Peasant Studies, 49*(6), 1139–1157.

Mustonen, T., et al. (2022). The role of indigenous knowledge and local knowledge in understanding and adapting to climate change. In H. Portner, et al. (Eds.), *IPCC climate change 2022: Impacts, adaptation and vulnerability* (pp. 2713–2807). Cambridge University Press.

Mosse, D. (2014). Knowledge as relational: Reflections on knowledge in international development. *Forum for Development, 41*(3), 513–523.

Sumberg, J., & Thompson, J. (Eds.). (2012). *Contested agronomy: Agricultural research in a changing world*. Routledge.

Stone, G. (2007). Agricultural deskilling and the spread of genetically modified cotton in Warrangal. *Current Anthropology, 48*, 67–103.

Taylor, M., & Bhasme, S. (2019). The political ecology of rice intensification in south India: Putting SRI in its places. *Journal of Agrarian Change, 19*(1), 3–20.

Taylor, M., Bargout, R., & Bhasme, S. (2020). Situating political agronomy: The knowledge politics of hybrid rice in India and Uganda. *Development and Change, 52*(1), 168–191.

Taylor, M., & Bhasme, S. (2021). Between deficit rains and surplus popluations: The political ecology of a climate resilient village in south India. *Geoforum, 126*, 431–440.

Further Readings

Baker, Z., Law, T., Vardy, M., & Zehr, S. (2023). *Climate, science and society: A primer*. Routledge.

Flachs, A. (2019). *Cultivating knowledge: Biotechnology, sustainability and the human cost of cotton capitalism in India*. University of Arizona Press.

Kumar, R. (2016). *Rethinking revolutions: Soyabean, choupals and the changing countryside in central India*. Oxford University Press.

Taylor, M. (2016). *The political ecology of climate change adaptation*. Routledge.

Of Climate, Communities and Commons: A Gendered Perspective

Smriti Das

Abstract Much of development policy and planning, particularly when it comes to preparedness to tackle climate change, has been "gender blind." This requires immediate attention, as we stand threatened by both the short and long-term impacts of climate change and climate vulnerabilities. Gendered vulnerabilities, accentuated by interlinkages of physical factors to social relations and hierarchies, will manifest through disproportionate impact on women who are pre-disposed owing to their household provisioning roles. These impacts are likely to be higher with social and ecological disruptions that will affect community resilience. By social disruption, I indicate a breakdown in collective institutions, changing power relations and fragmentation of local knowledge systems. By ecological disruption, I indicate vanishing commons and other ecological risks like loss of biodiversity. A gendered analysis of adaptation responses reinforces the significance of women's everyday experiences and knowledge systems in coping with adversities. A multi-scalar analysis deems it extremely significant for meso (related institutions of the state and government policies) and micro-level institutions to be gender responsive and create democratic spaces for women's participation. At the local

S. Das (✉)
XLRI Delhi-NCR Campus, Jhajjar, Haryana, India
e-mail: smriti.das@xlri.ac.in

49
N. Bikkina and R. M. R. Turaga (eds.), *Climate Change Adaptation*,
https://doi.org/10.1007/978-981-97-1076-8_5

level, the embeddedness of gender power relations in social, institutional and cultural contexts calls for cultural departures with shifts in gender relations.

Keywords Gendered vulnerabilities · Power relations · Climate change · Community resilience · Social disruption

My opening remarks in this panel will try and mainstream communities and commons into the climate debate, albeit with a gendered lens. The question that one might ask is why communities and commons and what does it have to do with gender? To understand this, we will situate the problem in context briefly. There is enough evidence to indicate the direct and indirect impacts of climate change (IPCC, 2023). In the coming times, not too far, climate change will further intensify environmental processes at the landscape level, which means it will impact biodiversity, trigger changes in river flow and the frequency of extreme events such as cyclones, hurricanes, droughts and floods will be high, depending on different regional characteristics. We all know fairly well about the impact of these changes on populations. There have been sufficient warnings by the scientific community about impending threats of global warming, glacial retreats, floods and several more. We have seen some major impacts in the Himalayan range in India, for instance, through the disaster events in Kedarnath (2013) and Chamoli (2021), both in Uttarakhand, India. The threat is global. Every continent and country will be affected, but some will be affected more than others. Similarly, some sections will be affected more than others. So, the impact is not the same for everyone.

Climate stressors will affect the adaptive capacity of human and non-human populations, but the adaptive capacities are the consequence of many other economic, social and political factors. These include the condition of natural resources, ownership of assets and resources, poverty and inequality levels, health conditions, social networks, political vision, the robustness of institutions, condition of global markets, international relations and so much more (Adger, 2006; Cardona et al., 2012). So, there are internal factors and local factors and then there are a range of external factors that are affecting it. There is tremendous work, and much more is desirable to understand the vulnerabilities of the human population to various climate stressors and the adaptive capacities of different population groups.

Through today's talk, we will try and understand the vulnerabilities and resilience of communities with reference to climate change. When we speak of vulnerabilities, we go beyond technically binding them by imagining the magnitude of exposure and risk that we normally do to focus on marginalities. There could be multiple reasons for this, such as social and economic inequalities, gendered inequalities, environmental inequities, institutional processes limiting access to natural resources, social networks, commons and collectives, insecure tenure and so on. These marginalities limit the adaptive capacities or the abilities of communities to learn and adapt to new situations. Intersected with social differences, the vulnerabilities and impacts become more acute. When we posit these contextual or social constructions of vulnerability against the long-standing multi-scalar phenomena, reconciliation looks much more difficult and complex.

CLIMATE ADAPTATION AND INTERSECTIONALITY

I will try to unpack gendered vulnerabilities and establish the need for more gender-responsive policies and institutional designs. It might certainly be unsettling for communities and societies that are still deeply embedded in the structures of patriarchy. I recognize that there is a school of thought that problematizes the gendered framing of vulnerabilities, arguing that such framing ignores the gender-differentiated knowledge and capabilities of a core group of stakeholders. However, my idea is to identify and bring to center stage the aspects of marginality that have and should inform the discussions on climate adaptation. What do I mean by gendered vulnerability? Gendered vulnerability draws our attention to pre-existing social conditions that shape the experiences of different genders, and here I focus on women. It is not that women or men are inherently vulnerable to these events. Gendered experience of the same event means that women and men are socially conditioned and situated in positions where they feel and perceive the impacts differently (Agarwal, 1994; Arora-Jonsson, 2011; Erwin et al., 2021). Social relations—relations of power experienced through differential access to and control over social and economic resources shape these experiences. These experiences are embedded in local–regional political economy, and as we intersect gender with other access to social differences, it becomes more scathing (It is a huge volume of work on intersectionality that I am referring to (including Carr & Thompson, 2014; Crenshaw, 1991; Kaijser & Kronsell, 2014; Thompson-Hall et al., 2016)). For instance, if we intersect gender with caste, class, religion and ethnicity we realize that these differences render some groups more vulnerable (Goodrich, Udas, & Larrington-Spencer, 2019; Sultana, 2014).

There may also be communities that are exposed to both ecological and social marginalities together. For instance, hill communities are exposed to much higher risks of climate hazards and other ecological fragilities that are getting accentuated with climate change, and we are currently witnessing it. When with changing ecological conditions and in the absence of livelihood opportunities men migrate to cities, it becomes extremely difficult for women to manage in the absence of economic and political freedoms. In a study done by my students in 2019 in the remote, hilly parts of Uttarakhand, we found that almost all decisions at the household level were taken by the male members (husband or the father-in-law). A pattern could also be drawn by looking at ownership and use of productive assets, including land, access to and user adaptability of technology, access to institutional credit, access to markets, leadership patterns and decision-making. Even though many households were women-headed in the absence of male members, they perceived it as mere tokenism with little or no change in gender norms. Though they acknowledge that migration of men allowed them more freedom of mobility and social interactions, when livelihood intervention was designed in this area, it overlooked many of these aspects and remained far from the intended goal of women's empowerment (Sharma, 2019; Sinha, 2019).

Numerous studies across Asia, Africa, Latin and Central America have explained how differentiated power relations and access to resources and institutions determine the coping abilities and the capacity to recover from various stresses and risks, which can be climatic or non-climatic. Translating these multiple embedded realities to the everyday experience of climate-induced disasters, one can see the gendered impact. Women are not only predisposed to natural hazards but are overburdened with responsibilities when male members out-migrate for a long term under conditions of distress. There may be a potential opportunity for female members also to migrate and unfetter themselves from social norms. However, where conditions are unfavorable, there are equal risks of sexual exploitation, workforce discrimination and isolation, as shown in different studies. Goodrich, Prakash, and Udas (2019), in their study in the Hindu Kush Himalayas area, have also shown how long-term out-migration might trigger higher dropout rates of girls from formal education and increase gender-based violence and trafficking of girls. They also highlight how many poor households might further be pushed into marginal and unsafe locations by fast-paced development and growing urbanization. Thus, we see their chances of social disruptions are quite high, and the consequences are many, which again are gendered. Thus, the prepared-ness of disaster resilience, disaster mitigation and engagement with the issue of migration need to be sensitive to gender roles and gender power relations and based on enhancing capabilities rather than reinforcing the differences.

Traditional Knowledge, Women and Adaptation

The other significant aspect in the debate is the fact that despite the history of environmental mobilization by women and recognition by some of the ecological knowledge of women, much of the analyses render women's critical role in nature conservation as an outcome of their "nurturing role" or, "caring and loving nature," and that is how we dismiss a great deal of contribution of women to nature. It took long to recognize that women-led environmental movements could also stem from their political and economic marginality. Women's close interaction with nature to support household livelihood leaves them with a close appreciation of natural phenomena. Traditional knowledge/practices have enabled women to adapt to changing climatic conditions. Adaptation studies in the Pacific by McLeod et al., (2018) show how women revived traditional practices for managing drought, including drying and fermenting bread for food security. Such practices are there in many parts of Asia as well. At another location, traditional wells enabled them to access potable water and build new shallow wells, and all this was done through the knowledge that was there with women. While much research on traditional values and climate change has focused on how local weather understanding complements large-scale climate predictions, environmental policies or adaptation policies still have a long way to go in factoring local knowledge, diversities and the cultural context.

The point I bring to the fore is that the tendency to draw binaries when talking about men and women runs the risk of reinforcing static categorizations that undermine social struggles, contestations and new identity formations (Alaimo, 2009; Kaijser & Kronsell, 2014). Gender is a social construct. As also shown by Goodrich, Udas and Larrington-Spencer (2019), in their work on the Hindu Kush Himalayas, gendered constructions are contested and renegotiated over time even in the same community. Arora-Jonsson (2011) articulates it quite interestingly by stating that a counterpart to women's vulnerability is their virtuousness. Such generalizations present a static conception of women's role and while women's vulnerability (to climate change in the current discussion) and virtue (in a caring role) can be portrayed to favor conservation outcomes, this formulation ignores the inequalities of power and how it may interplay with other factors to downplay the efforts of one group. While the inclusion of women is certainly desirable, its regressive outcomes in terms of reinforcing gender inequalities necessarily need to be guarded against. Inclusion of women, for instance, in forestry organizations (in India and Sweden), Jonsson states, was a status quoist measure because they were made to abide by rules without having many roles in framing these rules. Women preferred to participate in their own groups where they were more confident. This idea was rejected by male-dominated village/forest organizations that regarded women's agency and group formation as a challenge to their organizations. The point to take home is that institutional flexibility and adaptation are required to ensure that groups participate in decision-making and in making their own choices. The idea of retrofitting does not work too well with complex social and political institutions. Any change must be based on the recognition of women's role in environmental management and intra-household and intra-community decision-making.

Asserting Rights: The Case of the Women of Baiga Tribe

Nevertheless, there is more to it when we think of gendered participation in the environmental context. Where policies have been silent on gender or "gender blind" in terms of strategizing inclusivity and thus failed to improve environmental outcomes, women have mobilized and asserted their agency, and such instances are many. In a study by my doctoral student among the Baiga tribes of Madhya Pradesh, she witnessed how Baiga women had effectively mobilized against coupe felling and illegal extraction from the community-managed forest. The study helped to break the myth that indigenous communities in India are all homogenous and egalitarian; there were pieces of evidence of patrilineality and male supremacy in the decisions. Baiga women remain overburdened with reproductive and productive roles. They are also closely associated with forests, which cater to their livelihood needs like fuel wood or non-timber forest produce, etc. Despite their close association, their control over decisions remained weak. However, that did not deter Baiga women from resisting state control and coupe felling of their forest.

Starting in 2008, when a Baiga woman stood up against tree marking and felling by the people employed by the forest officials, several scuffles marked the encounter between the forest department and Baiga women till almost 2012. Baiga men, who initially dismissed their resistance, were gradually convinced by women to stand with them. After a long struggle, they managed to convince the forest department to withdraw from their forest, over which they had by now also got community title. The mobilization in the Baiga village gradually led to negotiated institutional spaces and recognition of their knowledge and contribution to the management of the forest. The village institutions created democratic spaces and demonstrated gender inclusivity. The existing community forest institutions (namely, the Forest Rights Committee and the Biodiversity Management Committee) were reconstituted to include women in their executive bodies. The institutions also ensured participation from female-headed households.

Realizing the constraints that women face in attending village meetings, the institutions modified rules to ensure women's participation in decision-making. For instance, the village meetings are now held in the evening, considering women's household chores and unavailability during the daytime because of their involvement in many other productive and reproductive activities. Gradually, there is a ripple effect, where more institutions are collectivizing to demand their rights and designing institutions for forest management (Tyagi & Das, 2020). The case study should also be seen as an example of how gendered environmental politics could lead to not only balanced conservation and livelihood outcomes but also challenging patriarchal norms, which come in the way of taking this knowledge forward. How well these institutions address issues of social/gender and environmental equity will determine their sustainability. However, for now, certainly, such examples give hope in the changing political and environmental context.

Conclusions

To conclude, I would like to slightly harp on environmental and development policymaking. Much of development policy and planning, particularly when it comes to preparedness to tackle climate change, has been "gender blind." Even if we talk about climate negotiations, gender as a category came much later in the agenda and functioning of the UN Framework Convention on Climate Change. Of course, it does factor in the Nationally Determined Contributions in most countries now. This (gender equity/inclusion) requires immediate attention, as we stand threatened by the short- and long-term impacts of climate change and climate vulnerabilities. Gendered vulnerabilities accentuated by interlinkages of physical factors to social relations and hierarchies will manifest through disproportionate impact on women who are predisposed owing to their household provisioning roles. These impacts are likely to be higher with social and ecological disruptions that affect community resilience. By disruption, I have indicated a breakdown in collective institutions, changing power relations and fragmentation of local knowledge systems. By ecological disruption, I indicate vanishing commons and other ecological risks like loss of biodiversity. A gendered analysis of adaptation responses reinforces the significance of women's everyday experiences and knowledge systems in coping with such adversities. A multi-scalar analysis is extremely significant for meso (related institutions of the state and government policies) and micro-level institutions to be gender responsive and create democratic spaces for women's participation. At the local level, the embeddedness of gender power relations in social, institutional and cultural contexts calls for cultural departures with shifts in gender relations. We have many case studies where this is happening, but the process is prolonged. In situating the gendered narrative, I have focused more on the experiences in the rural environmental setups. It holds also for urban contexts, more so as projections indicate that much of the population will be in the urban areas. The vulnerabilities are going to be much higher, and we need to rise and respond to the common call.

REFERENCES

Adger, W. N. (2006). Vulnerability. *Global Environmental Change, 16*(3), 268–281.

Agarwal, B. (1994). *A field of one's own: Gender and land rights in South Asia.* Cambridge University Press.

Alaimo, S. (2009). Insurgent vulnerability and the carbon footprint of gender. *Women, Gender, and Research, 18*(3–4), 22–35.

Erwin, A., Ma, Z., Popovici, R., Salas O'Brien, E. P., Laura, Z., Eliseo, Z. Z., Bauchet, J., Ramirez Calderón, N., & Roberto, A. L. G. (2021). Intersectionality shapes adaptation to social-ecological change. *World Development, 138*(2021), 105282.

Arora-Jonsson, S. (2011). Virtue and vulnerability: Discourses on women, gender, and climate. *Global Environmental Change, 21*(2), 744–751.

Cardona, O. D., Van Aalst, M. K., Birkmann, J., Fordham, M., McGregor, G., Rosa, P., Pulwarty, R. S., Lisa, E., Schipper, F., & Sinh, B. T. (2012). Determinants of risk: Exposure and vulnerability. In C. B. Field, V. Barros, T. F. Stocker, D. Qin, D. J. Dokken, K. L. Ebi, M. D. Mastrandrea, K. J. Mach, G. K. Plattner, S. K. Allen, M. Tignor, & P. M. Midgley (Eds.). *Managing the risks of extreme events and disasters to advance climate change adaptation: A special report of working groups I and II of the Intergovernmental Panel on Climate Change* (pp. 65–108). Cambridge University Press.

Carr, E. R., & Thompson, M. C. (2014). Gender and climate change adaptation in Agrarian settings: Current thinking, new directions and research frontiers. *Geography Compass, 8*(3), 182–197.

Crenshaw, K. (1991). Mapping the margins: Intersectionality, identity politics, and violence against women of color. *Stanford Law Review, 43*(6), 1241–1299.

Goodrich, C. G., Prakash, A., & Udas, P. B. (2019a). Gendered vulnerability and adaptation in Hindu-Kush Himalayas: Research insights. *Environmental Development, 31*, 1–8.

Goodrich, C. G., Udas, P. B., & Larrington-Spencer, H. (2019b). Conceptualizing gendered vulnerability to climate change in the Hindukush Himalaya: Contextual conditions and drivers of change. *Environmental Development, 31*, 9–18.

IPCC. (2023). *AR6 synthesis report: Climate change 2023.* Author.

Kaijser, A., & Kronsell, A. (2014). Climate change through the lens of intersectionality. *Environmental Politics, 23*(3), 417–433.

McLeod, E., et al. (2018). Raising the voices of Pacific Island women to inform climate adaptation policies. *Marine Policy, 93*, 178–185.

Sultana, R. (2014). Gendering climate change: Geographical insights. *The Professional Geographer, 66*(3), 372–381.

Thompson-Hall, M., Carr, E. R., & Pascual, U. (2016). Enhancing and expanding intersectional research for climate change adaptation in agrarian settings. *Ambio, 45*(3), 373–382.

Tyagi, N., & Das, S. (2020). Standing up for forest: A case study on Baiga women's mobilization in community governed forests in Central India. *Ecological Economics, 178,* 106812.

Findings from the fieldwork of the following Master's thesis from TERI-SAS have been mentioned in the talk:

Sharma, P. (2019). Evolving role of women in the context of male out-migration: A study in a hill district of Uttarakhand.

Sinha, I. (2019). Understanding women's well-being using the Capabilities approach.

Further Readings

Dengler, C., & Lang, M. (2022). Commoning care: Feminist degrowth visions for a socio-ecological transformation. *Feminist Economics, 28*(1), 1–28.

Harcourt, W. (2017). Gender and sustainable livelihoods: Linking gendered experiences of environment, community and self. *Agriculture and Human Values, 34,* 1007–1019.

Perkins, P. E. (2019). Climate justice, commons, and degrowth. *Ecological Economics, 160,* 183–190.

Tribal Communities of the Northern Eastern Ghats: Forest Dependence in the Context of Changing Climate

Vikram Aditya

Abstract India's Eastern Ghats mountain region, spread across Odisha, Andhra Pradesh, Telangana, Karnataka, and Tamil Nadu states, is predominantly inhabited by several forest-dwelling tribal communities who have distinct lifestyles and livelihoods linked to the forested landscape. Most of the Northern Eastern Ghats (NEG) is placed under the 5th Schedule [under Article 244(1) of the Constitution], which grants special provisions for the administration of Scheduled Tribes [Article 366(25)]. The major forest dwelling tribal communities in the NEG include the primarily hill-dwelling Konda Reddis, plains-dwelling Koya and Nayakpod cultivators, the recently settled Gutta Koyas (also known as Maria Gonds in Chhattisgarh), displaced from insurgency-affected Chhattisgarh and other groups such as the Gadaba, Bagatha, Savara, Valmiki, and Paraja tribes. These forest-dwelling communities live across the 'Agency' (Schedule 5 areas) hill tracts of East and West Godavari and Visakhapatnam districts of Andhra Pradesh. This study conducted close to a hundred and forty interviews with three tribal groups (Konda Reddys, Koyas, and Nayakpods) and other non-tribal

V. Aditya (✉)
Centre For Wildlife Studies, Bengaluru, India
e-mail: vikram.aditya@cwsindia.org

© The Author(s), under exclusive license to Springer Nature Singapore Pte Ltd. 2024
N. Bikkina and R. M. R. Turaga (eds.), *Climate Change Adaptation*,
https://doi.org/10.1007/978-981-97-1076-8_6

communities in the Papikonda NP, the sole National Park in the NEG of Andhra Pradesh, and its buffer forests over a three-year period from 2014–2017. Results from the interviews indicate a high dependence of these communities on forest resources, particularly through shifting cultivation, collection of minor forest produce and hunting that contributed to food and livelihood security. Community dependence varied with tribal groups and their location inside or outside the forest. Communities perceived significant forest degradation in the NEG over the past three decades, driven by plantations, agriculture, and dam building, and their observations of climate change were coupled with these land cover changes. I discuss how communities are adapting in the context of land cover change, climate change, forest management, forest rights and development schemes being implemented in recent years across the Northern Eastern Ghats.

Keywords National Park · Forest produce · Shifting cultivation · Forest dependence · Climate change

Papikonda National Park is the sole national park in the Northern Eastern Ghats and the buffer around it. I had talked to the local community members across the landscape to understand the extent of their dependence on the forest. I also had the opportunity to interact and ask them about their views on climate change.

Northern Eastern Ghats: Some Background

The Northern Eastern Ghats (NEG) landscape encompasses Andhra Pradesh and Odisha states of eastern and southern India. The Eastern Ghats is a hill region that is roughly parallel to the east coast of India. It stretches about 1600 kilometers long from north to south (Aditya & Ganesh, 2019; Goswami et al., 2018) (Fig. 1). The Eastern Ghats is also considered to be one of the oldest geological formations globally. Parts of the landscape are actually quite old. The rock formations of certain sections of the Eastern Ghats are about 3.5 billion years old (Pragasan, 2014). There are several different indigenous tribal communities that inhabit the landscape. The NEG is inhabited by communities such as the Koyas, Konda Reddys, Nayakpods, Kondhs, Sauras, Gadabas and

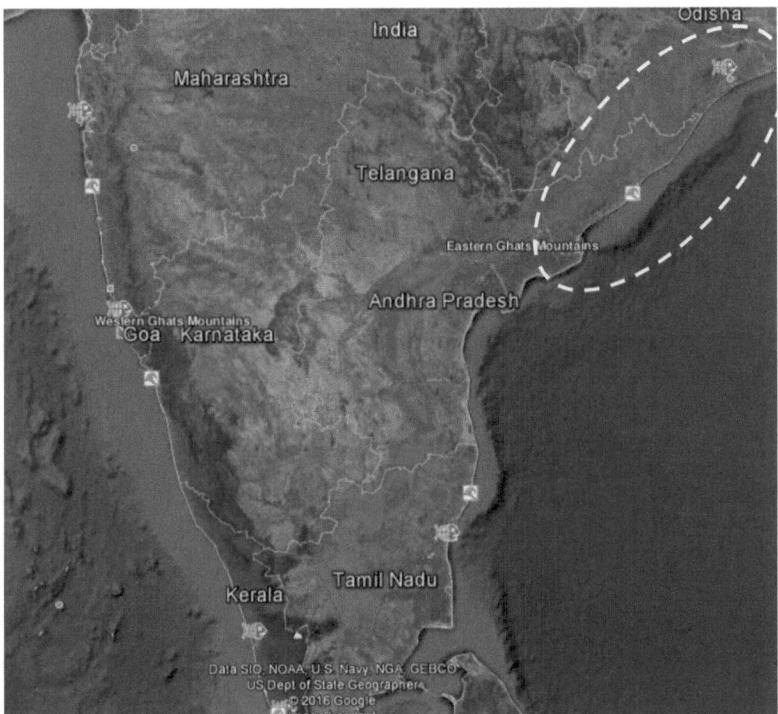

Fig. 1 Google Earth image of the Eastern Ghats, with the dashed line indicating the Northern section of the Eastern Ghats in Andhra Pradesh and Odisha

Bagatas (Aditya & Ganesh, 2019; Oskarsson, 2010). These communities reside forest areas both in Andhra Pradesh and Odisha. The Soligas are another forest-dwelling community in Tamil Nadu and Karnataka. The landscape supports diverse flora and fauna. We have identified 56 species of mammals in the Papikonda National Park as part of our research. Unfortunately, parts of the NEG have experienced decades of armed conflict between security forces. The NEG is also a region of intense and rapid landscape transformation (Aditya & Ganesh, 2019; Beehler et al., 1987; Ganesh et al., 2015; Srivastava, 2009). This is happening due to a number of factors. The main drivers of landscape transformation here are agriculture and dams, as well as plantations and shifting cultivation, which was practiced earlier, but not practiced anymore now.

The Godavari River flows from west to east bisecting the Papikonda National Park into nearly equal northern and southern halves. The Godavari works as a very important divide for a lot of species. Some species are only found in the north; while some are only found in the south. Papikonda National Park covers about 1012 square kilometers in area. The elevation ranges from about 20 meters at the banks of the Godavari River to about 800 meters at Bison Hill, which is the highest point in the region. The Park is located in the East and West Godavari districts in coastal Andhra Pradesh, and is one of the oldest protected areas in Andhra Pradesh. It was first declared as a wildlife sanctuary in 1978 and in December 2008 it was notified as a National Park. The forest type is predominantly moist deciduous. However, the lower reaches of the hills are covered with dry deciduous forests. We also see patches of semi-evergreen forest in the higher elevation ranges as you cross 700 meters.

DRIVERS OF LANDSCAPE TRANSFORMATION IN PAPIKONDA

This is a landscape that is experiencing intense transformation. If you look at the image of Papikonda NP and its buffer in 1991, a large area is covered with high-density forest with some patches, which are also forested vegetation but not high-density forest. If you see the image in 2014, in a period of two decades, a lot of area has been converted from a high-density forest to a low-density forest, indicating the scale at which forest conversion and degradation is occurring in the Papikonda National Park and more generally in the NEG landscape (Figs. 2a and 2b). The number of drivers causing this transformation is changing. A number of dams are being built. A minor irrigation dam called the Bhupathipalem Dam was built in 2014 in the buffer of PNP. This has caused a lot of deforestation and also displaced three villages. A state highway, which was a one-lane road earlier, was made a two-lane highway in 2014. A number of pipelines are being laid for supplying water. Since the Godavari flows through the landscape, a number of lift-irrigation schemes have also come up on the Godavari. These pipelines are being laid through the forest, to transport the water of the Godavari to the villages and towns, obviously causing quite a bit of deforestation. You also see remnants of fire as many forest plantations have come up. Once an area is demarcated for

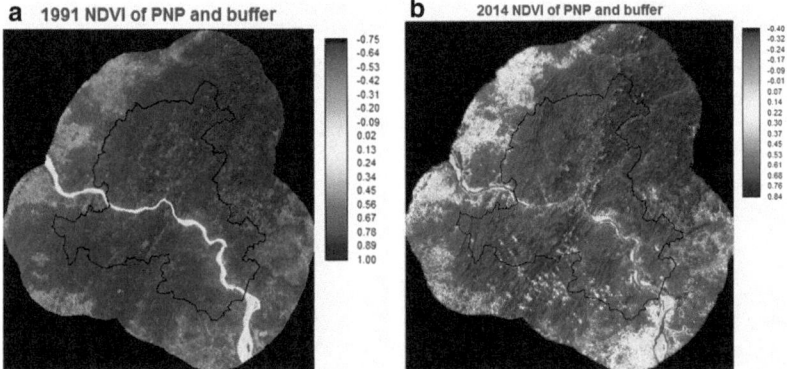

Figs. 2a and 2b Forest change in Papikonda National Park and its buffer between 1991 (left) and 2014 (right). Green indicates forest and yellow indicates non-forest area

plantation, that area is cut down and set afire to clear the ground vegetation to prepare the area for planting trees such as teak. This also obviously has caused quite a bit of land cover change in that region. There are coffee plantations near Maredumilli, which is one of the main villages just outside the national park. This has been there since 1996. There is also a plantation under Compensatory Afforestation Fund Management and Planning (CAMPA). This plantation was raised in lieu of land that was going to be submerged for the Polavaram Dam.

The Polavaram Dam is a large multi-purpose dam that is being built on the Godavari River, right adjacent to the Papikonda National Park; in fact, just abutting the boundary of the Park. Officially, the project is called the Indira Sagar Multipurpose Project as has been accorded National Project status by the Government of India. As one can probably imagine, this is going to cause a lot of submergence in villages, other habitats, farmlands, forests, etc. Within the National Park, it is going to cause the submergence of approximately 3267 hectares or close to four square kilometers of forest when completed. But outside the Park, in Andhra Pradesh, Chhattisgarh, and Odisha, it is going to cumulatively submerge roughly 60,000 hectares of forest in the NEG landscape (Fig. 3). Besides forest, large areas of farmlands and villages will also be affected.

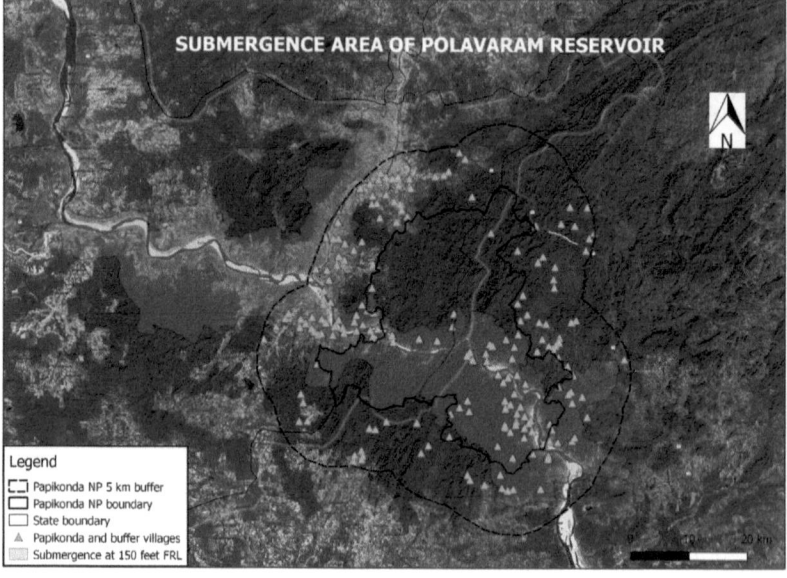

Fig. 3 Submergence area of the Polavaram project and villages affected (adapted from Naik et al., 2011)

The total estimated submergence area from Polavaram Dam is roughly 586 square kilometers. Besdies dams, *Podu* or shifting cultivation, which was historically practiced in the landscape, has caused the deforestation of about 69.13 square kilometers of forest (Figs. 4a and 4b). Plantations have converted 42 square kilometers of forest. Frequent fires have deforested 16.05 square kilometers of forest. The spillway of the Polavaram Dam itself has caused the conversion of 3.8 square kilometers. There have been five other minor irrigation dams and small dams that have come up in and around Papikonda, in the past three decades from the 90 s onwards. These have cumulatively caused the conversion of 0.14 square kilometers of forest area.

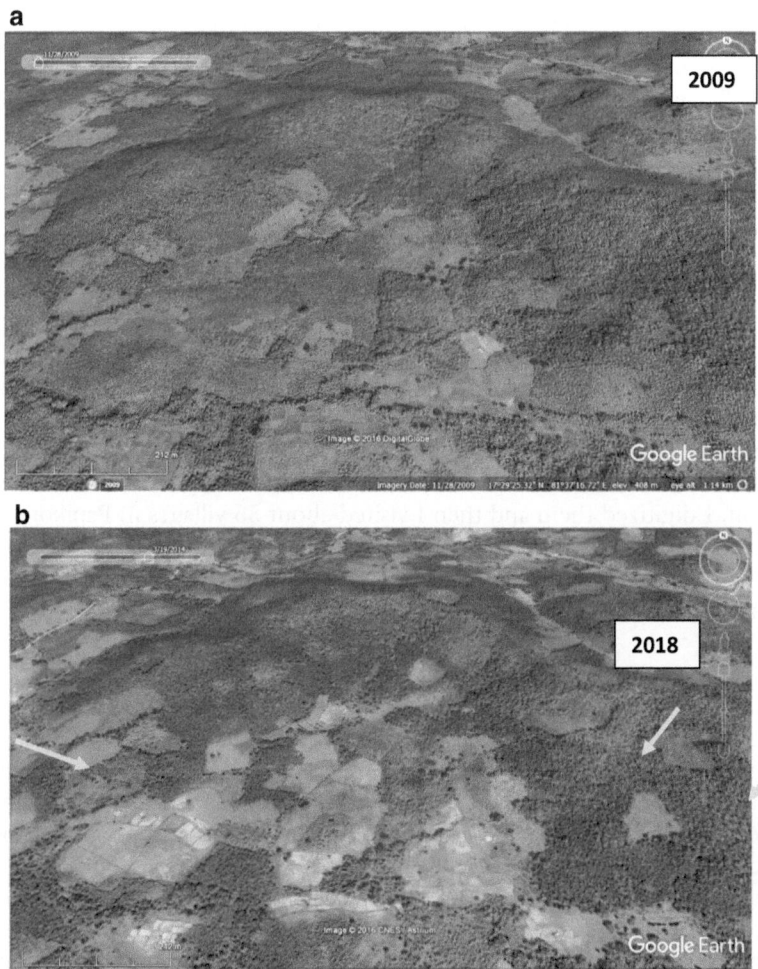

Figs. 4a and 4b Freshly cleared podus between 2009 (4a above) and 2018 (4b below) in the Northern Eastern Ghats

STUDY DATA AND METHODS

In this context, I interviewed communities to understand their perception of changes in the NEG landscape and their perception of changes in climate that have occurred in and around their villages and within their communities in the past three decades. I wanted to also understand how these changes are affecting the communities. I analyzed how these observations vary between tribal groups based on their location in the forest. I interviewed up to 140–150 members of different tribal communities. These included Koyas, Konda Reddis, and Nayakpods. These are the three main tribal groups inhabiting the landscape within the Papikonda NP region. I conducted a detailed GIS analysis to select the villages. I had also included the buffer villages because I wanted to compare and contrast not just community perceptions but also other ecological variables that I have studied in this landscape. Many of these villages are along the Godavari River. There are 212 villages roughly in the entire region. I digitized them and then I visited about 55 villages in Papikonda NP and its buffer and I finally selected 33 villages. These villages were all located at varying distances from the forest. I conducted a total of 138 interviews in these 33 villages.

PRELIMINARY RESULTS

First, I found that the level and extent of forest dependence was much higher among the Koya, the Konda Reddi, and the Nayakpod tribal communities inside Papikonda, even though there is the same amount of forest in the buffer also. The density of the forest is also much higher inside Papikonda. Communities depend on the forest for bamboo collection; primarily they use bamboo for a number of purposes. They make fences with it; they also make wooden stepladders, and kitchen instruments. So bamboo is absolutely vital for the tribal communities in the Eastern Ghats. They also collect a number of other minor forest produce like shrubs, fruits, etc. They also collect firewood. Gum collection and honey collection have been affected in the past two decades.

Second, I explored how climate change is manifesting in the region, according to the communities. Communities have reported an increase in the frequency, intensity and duration of forest fires within the landscape. Forest fires, they say, happen for a longer time now and they also happen much more frequently than a couple of decades ago. There is also a decrease in the duration of rain but an increase in the frequency of storms and extreme weather events. They say this is also affecting agriculture, because quite a bit of agriculture, especially in the hills happens to be rainfed and does not depend on irrigation. Climate change induced water stress also manifests itself in the form of drying up of bore wells, canals, and check dams. There are more frequent and intense floods that happen. Every year the level of the Godavari rises during July and August during the peak of the monsoon showers. However, according to the forest dwelling communities, there has been more of a catastrophic rise that is causing floods in the past two decades. When there is a very large flood, since many of the villages are located on the Godavari River itself, many of them are displaced. Their properties are affected, their homes are affected and they have to go up into the hills to save themselves. They say that this has happened more frequently over the past several years.

They also say that large trees in the forest, particularly Adina cordifolia, Pterocarpus marsupium, Terminalia tomentosa, Olea dioecea, etc., are not rejuvenating themselves successfully because of a lack of timely and sufficient rains. The majority of the community members also reported a decrease in tree density, mostly outside the Papikonda National Park, in the buffer region. This affects them in different ways because they depend on large trees for a number of livelihood-based activities. For example, they require honeycombs, which are only found on large trees, to extract honey. Earlier a lot of communities used to collect honey. But now they say that honey is not available so much because the large trees are missing or they are not present in such large numbers anymore. They are not able to get honey that easily. They also say that there is less fodder for livestock in the forest because of the lack of rain. There is not as much grass as there used to be before. They are also getting less forest produce like firewood and timber. Some minor forest produce yielding trees like soap nut and gum and also some fruiting trees have decreased. Another way in which this is manifesting itself is in the less availability of water in the forests.

They say that since there is less water available in the forest now than it used to be earlier, animals are coming down towards villages, particularly conflict animals like wild boar. Approximately 75% of the community members that I interviewed said that the frequency of wild animal sightings has reduced, whereas only the numbers of wild pigs and peafowls have increased. This is increasing crop raiding and there are more instances of the animals entering into conflict with humans. This is also resulting in more instances of hunting of wild animals (Aditya et al. 2021). They are also saying that plantations are drying up more frequently. The new plantations that they are raising, of different varieties of Eucalyptus, teak, etc., are failing because there is not enough rain like there used to be before or it is not occurring in a timely fashion. All of this is affecting their livelihoods significantly, because of which many villages now prefer to depend more on the national employment guarantee scheme (MGNREGS) and farm labor for employment, than to solely depend on forest for their food security.

CONCLUSION

To conclude, communities perceive that climate change is resulting in changes in the landscape, including a reduction of resources available in the forest for various purposes. They particularly report that large trees are missing now, because of which communities have to travel greater distances to harvest different produce like bamboo and other forest produce. There is a loss of gum, honey, and fruit-yielding trees. There are also more drought periods and extreme rainfall events. All of this is causing them to depend increasingly on employment guarantee schemes and farm labor.

REFERENCES

Aditya, V., Komanduri, K. P., Subhedar, R., & Ganesh, T. (2021). Integrating camera traps and community knowledge to assess the status of the Indian pangolin Manis crassicaudata in the Eastern Ghats, India. *Oryx, 55*(5), 677–683.

Aditya, V., & Ganesh, T. (2019). Deciphering forest change: Linking satellite-based forest cover change and community perceptions in a threatened landscape in India. *Ambio, 48*(7), 790–800.

Beehler, B. M., Raju, K. K., & Ali, S. (1987). Avian use of man-disturbed forest habitats in the Eastern Ghats, India. *Ibis, 129*, 197–211.

Ganesh, T., Vikram, A., Giridhar, M., & Prashanth, M. B. (2015). The 'empty forests' of the northern Eastern Ghats. *Current Science, 109*(3), 398–399.

Goswami, R., Thorat, O., Aditya, V., & Karimbumkara, S. N. (2018). A preliminary checklist of butterflies from the northern Eastern Ghats with notes on new and significant species records including three new reports for peninsular India. *Journal of Threatened Taxa, 10*(13), 12769–12791.

Naik, D. R., Bosukonda, S., & Mrutyunjayareddy, K. (2011). Reservoir impact on land use/land cover and infrastructure—A case study on Polavaram Project. *Journal of Indian Society for Remote Sensing, 39*(2), 271–278.

Oskarsson, P. (2010). *The law of the land contested: Bauxite mining in tribal, central India in an age of economic reform.* Unpublished Doctoral Dissertation, Norwich: University of East Anglia.

Pragasan, L. A. (2014). Carbon stock assessment in the vegetation of the Chitteri Reserve Forest of the Eastern Ghats in India based on non-destructive method using tree inventory data. *Journal of Earth Science & Climatic Change, S11,* 001. https://doi.org/10.4172/2157-7617.S11-001

Srivastava, V. (2009). Need for conservation of biodiversity in Araku Valley, Andhra Pradesh. *Current Science, 96*(1), 11.

Forest Rights and Climate Change: Conservation Through Indigenous Knowledge

Bijayashree Satpathy

Abstract Trees are the major weapons in fighting global warming and so are the forests. Forest has the potential to absorb approximately a tenth of carbon emissions worldwide. But once disturbed, forests contribute around a sixth of carbon emissions globally. Signatories of the 2015 Climate Accord in Paris were to work towards cutting emissions from the use of fossil fuels and encouraging the protection of forests. The local forest dwelling communities and their indigenous knowledge have been playing a vital role in conserving forest resources. Sacred groves are examples of the conservation of forests by the native communities in situ. India is believed to have the highest concentration of sacred groves, which have socio-cultural, ecological and economic significance. However, currently, the size and number of sacred groves are decelerating at an alarming rate due to industrialization, infrastructure development and agriculture expansion. In this backdrop, the study is conducted in two tribal villages each in Mayurbhanj and Keonjhar districts in Odisha, which are chosen purposively to understand the role and function of sacred groves in the

---

B. Satpathy (✉)
Agzistence Intelligence Private Limited, Mumbai, India
e-mail: Dr_Bijayashree@agzistence.com
URL: https://agzistence.com/

N. Bikkina and R. M. R. Turaga (eds.), *Climate Change Adaptation*,
https://doi.org/10.1007/978-981-97-1076-8_7

lives of the forest dwelling communities. Mining and dam construction are some of the factors that have affected sacred groves adversely. It was found that the area under sacred groves is reduced and the dense groves are turning sparse and bare. The forest dwelling communities through their indigenous practices collectively conserve these groves. However, the rights of the tribal communities inhabiting these areas are unrecognized under the Forest Rights Act as of now. Rights recognition is necessary to encourage the collective action of the community conservation initiative for sustainable conservation.

Keywords Sacred groves · Forest rights · Protected areas · Common property rights · Tribal communities

As you venture into Similipal Tiger Reserve, you may encounter the awe-inspiring Heritage Tree Sals. Having conducted fieldwork in both the core and buffer areas of wildlife sanctuaries, I have observed that these protected areas are vital for conserving forest and wildlife. However, within the protected areas, conservation of certain habitats are often overlooked. One such overlooked forest patch is a 'sacred grove' (Bhagwat & Rutte, 2006). It was found that there are limited provisions, both in the prevailing forest conservation legislation in India and in the existing global protected area network, with reference to sacred groves, specifically.

For generations, local forest-dwelling tribal communities have upheld a legacy of preserving their precious forests through their indigenous practices and customary institutions. However, the unfortunate reality is that multiple external communities have been exploiting these forests heavily. During the British rule in India, forests were initially cleared, and a forest policy was later introduced to advocate for forest protection by curbing traditional practices of the forest-dwelling tribal communities. The colonial administration implemented a system of surveys and geographical mapping of the forests to exert control over these areas (Kumar and Kerr, 2013), ignoring the common property rights and traditional practices of the indigenous communities. This conduct was an attempt by the colonial state to develop forest regimes that required physical expansion, which devalued the existing tenure systems and rights of forest dwelling tribal communities over these forest areas (Bhaduri, 2020).

Forest mapping in India's colonial era formed the basis of scientific forestry (Guha and Gadgil, 1988) in the country with the objective of reversing forest degradation, yet it came at a cost to tribal communities whose rights were restricted. The government sought to control forests for timber, leading to the conversion of the mixed woodlands of the Western Ghats into teak and those in Himalayan region to coniferous forests. The colonial administration also replaced many useful species cultivated by the forest dwelling tribal communities with commercial plantations. Despite India's Independence, most forest conservation policies, such as Joint/Community Forest Management and Social Forestry, advocated the dessicationist approach. The notion of Green Colonialism emerged, advocating for forest conservation to adapt to climate change while allowing the state to monopolise the control over these landscapes.

In the context of India's conservation and development efforts, conflicts have emerged between conservationists, preservationists, and indigenous communities due to their different perspectives on the use of forest resources. The postcolonial history of Indian conservation and development has witnessed the emergence of green nationalism, replacing the earlier green colonialism (Kumar, 2010). In the 1970s, displacement of indigenous communities became a norm in the name of conserving protected areas, leading to the exploitation of wildlife. The postcolonial history of Indian conservation and development sets the tone for the current conservation issues in India. Currently, conservation efforts in India prioritise human-free wilderness areas and tourist-friendly wildlife theme parks, where the urban elite can seek entertainment through wildlife watching.

Simultaneously development projects like mining, construction and other logistics and networking projects often clash with wildlife conservation efforts. Wildlife habitat was a casualty in the process. Nationwide welfare programs have been introduced to provide basic needs for communities living below the poverty line in rural India. But have they delivered on these promises?

Millions of Indians including forest dwelling tribal communities who live below the poverty line also remain beyond the radar of these development programs. Approximately 4.3 million population, including indigenous communities living in the protected areas, depend on forest resources, partially or fully for their sustenance (Kukreti, 2020). Preservationist and wildlife conservationist on one hand blamed the indigenous communities for their dependence on forest resources, especially for their poor consumption patterns or use of resources like slash and burn cultivation, farm overgrazing, hunting, etc., and on the other, there are those advocates of environmentalism of the poor who often view that the poor are on the side of conservation of forest resources as their livelihoods depend upon them. So the entire discourse is to conserve forests on the basis on its utilitarian objective, but it is necessary to recognise that forests have cultural and spiritual significance for indigenous communities. In a democratic setup, social and economic inequities leave very limited scope for negotiation between the state and community institutions to use, access and claim rights over forest resources while ensuring their sustainable use.

SACRED GROVES, CULTURE AND FOREST CONSERVATION

It is crucial to comprehending the reasons why forests conservation is necessary. The ongoing debate revolves around whether we should preserve forest resources to meet development needs or if forests should be safeguarded for cultural reasons. In this context, it becomes necessary to understand the significance of sacred groves in the lives of the forest dwelling tribal communities. My research utilised methods such as oral history and semi-structured in-depth interviews.

Sacred groves are present in almost all states of India, including Odisha. These groves are conserved through informal institutions based on beliefs and taboos since the dawn of human civilisation. The informal institution of sacred groves in Odisha is recognized by various names like *Jahira*, *Thakuramma* among others. Social fencing by local communities usually protects these groves along with their diversity of species. Studies in Odisha show a wide distribution of sacred groves varying in size, numbers and ownership (Rath & Ormsby, 2020). Many tribal communities practice clan-based management restraining themselves from entering into sacred groves and collecting fuelwood, logs and other non-timber forest products. This makes the sacred groves rich depositories of medicinal and aromatic plants, wild crops and many significant floral and faunal species.

The selection of two villages in Mayurbhanj and Keonjhar districts was purposeful for the purpose of this study. One village is situated near Simlipal, while the other lies near the Jharkhand border. In Keonjhar,

Joda is the area where intensive mining activities are taking place, and this study purposively selected two villages within close proximity to the mining areas. One of the studied villages, Kumudabadi, in Mayurbhanj, is inhabited by multiple tribal groups, each with its own *Jahira* or sacred grove. The remaining studied villages own a Jahira for the entire village.

The customary conservation practice of sacred groves has evolved as an informal institution from endogenous socio-cultural codes. This has resulted in socio-cultural practices that assign roles to participants and guide interaction among forest users. The forest dwelling community delineated the territory by social fencing. The communities celebrate festivals and make decisions regarding the conservation of sacred grove collectively through open interaction. However, the participation of women in decision-making concerning management of sacred groves is limited. Women rarely play a role, and their entry to the sacred grove is prohibited, even when they are allowed to attend festivals outside the boundaries of these territories. The deity is present in the middle of the sacred grove, which is full of Sal trees. People offer domesticated goats and fowls to the deity.

The position of the priest of the Jahira is inherited by the descendants of a particular household and the decision of when and how to celebrate the festivals lies with him. The priest stated "...*this is our Jahira sala...they are divine souls, standing there for centuries. No one is allowed to step into the grove, forget about tree felling. Harming the grove invites bad omen and diseases.*" Here we can see communities' sense of ownership coupled with the strong belief system. The Hill Kharia community in the village is the most vulnerable. They have faced multiple displacements, including displacement and landlessness during the construction of the dam in the early 1980s. Prior to that, they were displaced from the core area of Simlipal Tiger Reserve. Once they left the area, they had to leave their *Jahira* too. Many sacred groves were destroyed during the construction of the dam near the studied village. Similarly, in the villages in Keonjhar, the *Jahira* or sacred groves are destroyed due to various mining activities.

The Forest Rights Act 2006, Section 3.1(I), enables indigenous forest dwelling communities with the ability to protect and conserve their forest resources. However, a study of villages reveals that the sacred groves, though established as community-owned mechanisms to cope with the shocks of nature, remain unrecognised and unclaimed. The degradation of Sal trees, a predominant species in *Jahira*, has been a source of concern for the local communities, who view conservation as a necessary task. One of the villagers shared his concerns that *"…few old Sal trees are left inside the Jahira. There is no boundary to protect it. We need a cement-concrete boundary of around three feet to protect our Goddess from the untamed animals. We want some support for that from the Tribal and Forest Departments."* To address these concerns, the State initiated the Integrated Development of Sacred Groves (IDSG) initiative in 2013, which has identified 2600 sacred groves. These groves are primarily located in areas inhabited by tribal groups that comprise 23% or 9.6 million of the state's 40 million population. It is noteworthy that no sacred grove in mining areas of Keonjhar has been demarcated or provided assistance. In the mining areas, the communities has lost sacred groves, which were previously used to cope with the impacts of nature.

Communities are susceptible to unforeseen impacts of nature and sacred groves are community-designed mechanisms to cope with such shocks. Sacred groves bind communities together, fulfilling a psychological and social requirement. They are typically owned by clans or villages and have taboos associated with them that act as a tool for monitoring the misuse of the grove. Festivals and religious ceremonies provide opportunities for frequent community interaction, which becomes a platform for information sharing and assessment of the groves' status and condition. The sacred groves are also symbolic of justice to the tribal communities, as these territories host tribal courts that resolve disputes among their members. The local communities' customary dependence on sacred groves helps to use the forest resources as sources of power, empowering the local communities to bargain with the forest department for the conservation and protection activity of sacred groves.

THE ECONOMICS AND POLITICS OF SACRED GROVE CONSERVATION

During the initial phase of the Integrated Development of Sacred Groves in 2013, approximately one lakh rupees were allocated for the initiation of plantation drives in 500 sacred groves. In 2015, the Odisha government launched a program that aimed to identify and demarcate sacred groves, ensure their maintenance for three years, prepare management plans for conservation, and raise awareness about their significance. In some sacred groves, concrete walls of two to two-and-a-half feet are built. But approach roads to some sacred groves are yet to be laid. Forest officials attempted to implement the State government's Integrated Development of Sacred Groves program. Forest officials are entering into the sacred groves, clearing the undergrowth in the sacred groves and dumping bricks and other construction material inside them. As outsiders entry into the sacred groves are prohibited, so the villagers protested and since then the forest officials have not returned to the villages. In yet another instance, forest officials constructed a shed over the forest deity, planted approximately one hundred deodar and mango trees without informing the village forest committee, took pictures and departed. Unfortunately, most of the trees did not survive.

The lack of consultation with tribal communities, who are the primary guardians of sacred groves, is a failure on the part of the government in designing the program. Despite a history of limited success, the Joint Forest Management program committees are the sole planners of all interventions. The committee will decide on the allocation of the one lakh rupees grant under the program. It seems the historical rent seeking power of state officials may still be hesitant to relinquish their historical rent seeking power on forest resources.

CONCLUSIONS

The concept of sacred groves, which were initially intended for non-human use, has become increasingly valuable over time due to its utility owing to their presence in forested areas. In the Indian state of Odisha, these sacred groves are an integral way of life for the 9.6 million tribal inhabitants who occupy the region (Sahoo, 2015). Sacred groves are considered holy and are strictly off-limits to human intervention, especially from outsiders, as it is believed that any disturbance could offend the deity responsible for their upkeep and trigger death, disease or catastrophe. These groves serve as socio-cultural hubs and provide a heaven for untapped biodiversity. Cross-scale environmental experts acknowledge that these sacred groves are rich repositories of undisturbed biodiversity hosting wild species of flora and fauna, coexisting in a symbiotic manner. This is the best example of *in situ* conservation. They are home to numerous indigenous medicinal flora, which are extinct otherwise, survive within these groves. The vegetation in the sacred groves helps to preserve the topsoil, while also improving soil fertility. The species surrounding the water bodies within these groves also play a critical role in maintaining water cycles in these landscapes. It is vital to recognise the forest rights and ownership of these sacred patches by the local community under FRA and left to them for their conservation and management. The changing climate also makes it pertinent to conserve such forest patches within the forested areas.

While cross-sectoral expertise in climate economy continues to emerge from various disciplines, it is crucial to acknowledge that indigenous forest inhabitants have long held a profound comprehension of organic economics. Their employment of sacred groves is a prime illustration of their depth of sustainability wisdom that these communities tradition- ally hold and practice through a frugal use of forest resources through economical tactics. This wisdom of the native communities manifested through the intergenerational ritualistic practices and customary beliefs, unacknowledged by the scientific community. Although the scientific sector confronts obstacles in comprehending and executing sustainable practices, these communities have valuable lessons to impart regarding sustainability.

References

Bhaduri, A. (2020, October 2). Poor implementation of forest rights act hurts tribals. *India Water Portal*. Retrieved on September 21, 2023, from the World Wide Web: https://www.indiawaterportal.org/articles/poor-implementation-forest-rights-act-hurts-tribals

Bhagwat, S. A., & Rutte, C. (2006). Sacred groves: Potential for biodiversity management. *Frontiers in Ecology and the Environment, 4*(10), 519–524.

Guha, R., & Gadgil, M. (1988). State forestry and social conflict in British India: A study in the ecological bases of agrarian protest. Bangalore: Indian Institute of Science.

Kukreti, I. (2020, October 5). 'Protect & Conserve model' displaced 13,450 families from 26 protected areas in 2 decades. *Down to Earth*. Retrieved on September 21, 2023, from the World Wide Web: https://www.downto earth.org.in/news/wildlife-biodiversity/-protect-conserve-model-displaced-13-450-families-from-26-protected-areas-in-2-decades-73656

Kumar, V. R. (2010). Colonizing greens: Political economy of deforestation in colonial South India 1800–1900. *A Biannual Journal of South Asian Studies, 3*(1), 226–240.

Kumar, K., & Kerr, J. M. (2013). Territorialisation and marginalisation in the forested landscapes of Orissa, India. *Land Use Policy, 30*(1), 885–894.

Rath, S., & Ormsby, A. A. (2020). Conservation through traditional knowledge: A review of research on the sacred groves of Odisha, India. *Human Ecology, 48*(4), 455–463.

Sahoo, S. (2015, September 1). Sacred Flaws: Why the Odisha government's move to save sacred groves has enraged tribal communities. *Down to Earth*. Retrieved on September 21, 2023 from the World Wide Web: https://www.downtoearth.org.in/news/forests/sacred-flaws-50989

Further Readings

Gadgil, M., & Chandran, M. S. (1992). Sacred Groves. *India International Centre Quarterly, 19*(1/2), 183–187.

Gadgil, M., Berkes, F., & Folke, C. (1993). Indigenous knowledge for biodiversity conservation. *Ambio, 22*, 151–156.

Gadgil, M., Berkes, F., & Folke, C. (2021). Indigenous knowledge: From local to global: This article belongs to Ambio's 50th Anniversary Collection. Theme: Biodiversity Conservation. *Ambio, 50*(5), 967–969.

Ormsby, A. A. (2011). The impacts of global and national policy on the management and conservation of sacred groves of India. *Human Ecology, 39*(6), 783–793.

Sharma, S., & Kumar, R. (2021). Sacred groves of India: Repositories of a rich heritage and tools for biodiversity conservation. *Journal of Forestry Research, 32*, 899–916.

Integrating Decarbonization with Adaptation Measures

Suman Chandra

Abstract Climate change has touched everyone's life in some form or the other- be it the next-door trash problem,the intensifying heat waves, rapidly drying rivers, forest fires or the rising number of endangered flora-fauna one grew up with. There is a growing recognition of the fact that we are on the brink of planetary crisis. However, many praise-worthy efforts to tackle climate change are underway globally. But do we have time for these one-off, few and far in-between efforts? What will it take to stop the ticking bomb? Do we have a shared under-standing of what decarbonization and adaptation pathways are relevant in our context? What additional steps will be required to accelerate these pathways? An integrative approach to delivering decarbonization and adaptation solutions requires a focussed analysis!

Keywords Decarbonization · Climate change · Adaptation · Urbanization · Vulnerability mapping

S. Chandra (✉)
New Delhi, Delhi, India
e-mail: sumanchandra.ias@gmail.com

We are amongst the top four emitters of carbon globally. USA is the highest emitter, followed by China, and EU. India is most populous country with growing energy needs and therefore our emissions are constantly rising and will continue to rise in the coming decade. When we consider historical emissions, India has not contributed much and we form a very minuscule percentage. Additionally, even our current per capita emissions are very low, almost one fourth of the developed world average. But this must not make us complacent, because we are a growing economy and our emissions are constantly on the rise, which is a worrying factor climate-wise in a business-as-usual scenario. For comparison, the US emits about 16.1 tons of carbon per capita, while India emits 1.9 tons. We ranked as low as at 148 rank in the per capita carbon emissions annually at global level (Emissions Database for Global Atmospheric Research, 2021).

One cannot overstate the impact of the climate change. The recent devastation in Uttarakhand underscores a concerning trend: extreme weather events are becoming increasingly frequent and disruptive, impacting all facets of life across the globe. Nobody is untouched. and unaffected with these changes. Climate change casts a long shadow, affecting everyone in some way. Thankfully, the conversation is evolving. It's no longer just talk – it's becoming a serious discussion grounded in science and actionable policy solutions. The impact of climate change will be more pronounced as we witness rapid urbanization and the sustainability of our resources is put to test. It's impact will manifest and will be seen more in the kind of air we breathe, the ocean we swim in and the society we live. It is also going to manifest itself in public health hazards and in the changes and trade-offs of how we use our natural resources given our growing developmental needs. In this face of deteriorating climate thus, the crucial question remains: who will bear the cost of climate inaction? The cost is high and most of it is going to be borne by the vulnerable communities. Should we be doing anything to address the question? We have to start thinking through our policy, through our day-to-day living and through our lifestyles about what changes are necessary to address this challenge. These changes and corrective steps have to be accelerated rapidly because we do not have that luxury of time as we are standing at the brink.

URBANIZATION AND CARBON

When we talk about mitigation and adaptation, what are we really referring to here? To put broadly, climate mitigation requires focus on addressing emissions reduction measures sectorally, while at the same time adapting to changes which are already underway due to irreversible climate change that have already taken place. One tangible option is to address the growing energy needs issue and the larger land-use issue. Additionally it will help to pay attention to what decarbonization pathways we are taking in each emissions contributing sector. At the heart of this lies the critical challenge of managing humanity's impact on the environment, specifically focusing on controlling the emissions we release. We must consider what are we doing about aout our energy consumption needs and how we are weaning ourselves from our dependency on the fossil fuels. This needs answering the question about how we are switching to the renewable energy paradigm. Additionaly active consideration of advanced options like geo-engineering or carbon capture and sequestration available to decarbonize including green hydrogen as a responsible fuel for the future with a potential to solve the energy problem of hard-to-abate sector has to be aggressively advanced. These mitigation measures have to be complemented withadaptation measures These are various mitigation and adaptation pathways, but potential of urbanization to arrest climate change by integrating mitigation and adaptation is maximum in developing countries which are still at various stages of building new form of urbanization. India that we experience today is an aspirational India!India that we experience today is an India, that wants to have public amenities, wants to live in good cities, which is why we see the constant thronging to the cities and a an advancing of urban agglomeration and of urban slums. Rapid urbanization is an inevitable development paradigm which will overwhelm us in coming decades. The absence of strategic urban planning efforts risks uncoordinated and potentially detrimental urbanization patterns. A proactive approach, emphasizing well-defined plans, is therefore essential for navigating the challenges and opportunities presented by this global phenomenon, especially in countries like India.

Numbers help to put the conversation in a context. The global population in 1900 was around 0.7 billion and in 2018 it was 4.2 billion. By 2050, we are expecting that it is going to be around 6.7 billion.

We can clearly discern that the numbers are rapidly rising, the urbanization today is at an unprecedented speed and scale. Considering the case of India, national projections indicate an imminent transition to a majority urban population, with estimates suggesting a potential fifty percent urbanization rate within a short timeframe.(Jha, 2020). Current estimates suggest India's urbanization level is approaching 40-44%, indicating a near-even split between rural and urban populations. Projecting forward 5-10 years, this trend aligns with global forecasts predicting 6.7 billion people residing in urban areas by 2050. Notably, a significant portion of this urbanization will originate from countries like China, India, and specific South African nations. Collectively, these regions are projected to contribute nearly one-third of the global urban population by 2050, signifying a substantial demographic shift.

In the context of human geography, urbanization can be defined as the progressive increase in the proportion of a population residing in urban centers.But what does urbanization essentially mean? When a person goes from a village to a city, what does that mean? Why do we go to urban spaces? We go there first for the employment needs, we go there for the better amenities, we go there for a better lifestyle and other serviceswhich are not accessible in a rural areas like a better transportation structure, a better energy structure. Therefore, we can understand that urbanization is driven by a multitude of factors, including the pursuit of employment opportunities, access to improved amenities, and a generally enhanced quality of life. These factors, often unavailable in rural settings (e.g., superior transportation infrastructure, reliable energy provision), act as a magnet for population migration. Consequently, rapid urbanization fosters a surge in demand for construction activities, increased energy consumption, and heightened pressure on existing resource utilization. In the case of India, with its projected population peak in the mid-century, a crucial window of opportunity exists. This timeframe enables proactive urban planning strategies to be implemented, circumventing the pitfalls of uncoordinated urban sprawl, which is typically resource-inefficient.

Even though it is counterintuitive,cities are essentially more efficient than the villages. Why? The impending surge of urbanization necessitates a paradigm shift from organic urban sprawl towards meticulously planned development. This strategic approach presents opportunities to achieve economies of scale in transportation and energy infrastructure, fostering more efficient resource utilization. Compact urban design, a cornerstone of planned urbanization, minimizes energy flows within city limits. By optimizing this crucial variable, we can significantly mitigate the environmental impact of urbanization and contribute to broader climate change mitigation strategies. While the concept of planned urbanization alongside rapid population growth might seem counterintuitive, it presents a powerful tool for both adaptation to and mitigation of the environmental challenges associated with large-scale urbanization.

Even within one state in India, one can see stark disparities if we take for example Maharashtra.

CLIMATE CHANGE MITIGATION, ADAPTATION AND INCLUSIVE ACTIONS

We have to start looking at our adaptation measures from a holistic matrix paradigm. A holistic matrix paradigm is essential for framing effective climate change adaptation and mitigation strategies. This approach necessitates transitioning beyond emotionally charged rhetoric towards a systematic and data-driven approach to environmental challenges. Effective climate change adaptation necessitates a multi-pronged approach. Firstly, vulnerability mapping at the sub-national level is crucial for identifying areas most susceptible to climate impacts. Scientific data analysis plays a vital role in understanding the historical and projected climate trends within a specific region. Secondly, the question of climate finance cannot be overlooked. Identifying sustainable funding sources for adaptation measures requires careful consideration. Finally, fostering close collaboration with relevant stakeholders, including local communities, is paramount. Community engagement ensures that adaptation strategies are not only effective but also socially responsible.

Rather than having very top-down focus, we need to have a more bottom-up approach to how to build up adaptation measures and that would include both traditional and modern methods. Climate change adaptation strategies must acknowledge the value of existing, traditional ecosystem conservation practices, which may in some instances prove superior to top-down policy interventions. Additionally, fostering early public outreach and education is crucial. While broad public awareness of climate change exists, its specific ramifications are often overlooked. For instance, the tragic phenomenon of drought-induced farmer suicides in Maharashtra and the central belt of India is frequently attributed solely to loan burdens and financial hardship. However, a climate-focused analysis reveals the underlying factor to be the persistent failure of monsoons and the excessive dependence of agricultural systems on these seasonal rains. In this context, it becomes essential to prioritize adequate post-disaster compensation for affected farmers, alongside a transition towards more resilient agricultural practices, such as the implementation of rainfall insurance schemes.

References

Emissions Database for Global Atmospheric Research. (2021). Fossil CO_2 emissions of all world countries, 2020 report. *EDGAR—Emissions Database for Global Atmospheric Research*. Retrieved on September 21, 2023, from the World Wide Web: https://edgar.jrc.ec.europa.eu/emissions_reports

Jha, R. (2020, August 3). Census 2021: Likely confirmation of past urbanization trends. *Observer Research Foundation*. Retrieved on September 21, 2023, from the World Wide Web: https://www.orfonline.org/expert-speak/census-2021-likely-confirmation-of-past-urbanisation-trends/

Impact of Climate Change on Forest Dwelling Communities and the Role of Forest Department in Mitigation

A. Rama Mohana Reddy

Abstract Climate change is a recent phenomenon and, hence, is yet to be understood fully. Up to 70% of the world's animal and plant species rely on forests for survival. Human activity involved in infrastructure, commercial logging, cattle ranching, urbanization and mining are also the major contributors to climate change. At the current rate of deforestation 1–10% of Tropical rainforest species are lost per decade by extinction. The zero net deforestation policy in safeguarding climate change and forest laws will increase food, water security and biodiversity conservation. Climate change impacts forest dwellers by creating shortages of drinking water, water for cattle, shortage of fodder, invasion of alien weeds, unforeseen droughts and floods, changes in crop patterns, frequent forest fires and spread of epidemics. Forest Departments therefore play a key role in mitigating the impact of Climate change on the forest dwelling communities.

Keywords Logging · Forest fires · Forest dwelling communities · Climate change · Afforestation

A. Rama Mohana Reddy (✉)
Andhra Pradesh, Visakhapatnam, India
e-mail: armreddyifs@gmail.com

87

The universe came into existence 4600 million years ago and of course, the universe was flooded with oxygen, about 2000 million years back and the whole climate change issue started to arise 200 years back. We have got very limited history of climate and the first conference on climate change took place in 1979 in Geneva. Our knowledge of climate change is very limited. We have been observing it and we are experiencing it now.

I will discuss here briefly the recent warning bells of climate change, the history of climate change, the impact of climate change on forest dwelling communities, facilitating adaptation to climate change and climate change mitigation. A very recent reminder to us of the warning bells from climate change is that of the Uttarakhand glacier burst (Mashal & Kumar, 2021, February 8), where more than 200 people lost their lives due to the mudding of the glacier at Joshimath, which took away with it the Dhauliganga hydropower project. Interestingly, water has become such an important commodity that, it is being traded on Wall Street. Water has become a future commodity and is now being traded on par with gold.

THE VISIBLE IMPACTS OF CLIMATE CHANGE

Carbon dioxide levels are at present 440 PPM. 500 million years ago, it was 220 PPM. More than 1.5 billion tons of carbon dioxide is released into the atmosphere every year from the cutting and burning of the forests alone. Up to 70% of the world's animal species are dependent on forests for survival and at this current rate of deforestation, we are losing 1–10% of the tropical rainforest species every decade. For instance, we no longer see frogs tweaking, snails roaming and House sparrows chirping in our premises, like what we used to see 25 years back. These are all the warning bells that we are getting from nature about climate change.

Talking about the impacts of climate change, the first thing that attracts our attention, is the scarcity of water. There will be a great shortage of water and shortage of fodder for the cattle of the forest dwelling communities. While the British had constructed catchment areas, like Shimla Catchment, Asia's Densest Forest, with no biotic interference, no such serious efforts are made now in many parts of the country. We are sacrificing natural forests in the name of development. A number of Invasive Alien weeds are constantly invading the state of Himachal Pradesh and many parts of the country, with no exception of any part. An obnoxious weed called *Lantana camara* is spreading on millions of hectares of forest and community lands (Bhagwat et al., 2012). Untimely rains are leading to crop damage and less yield of apples. There is a change in the crop pattern. The habitats of animals are destroyed and even monkeys are invading the human habitations and with that, they are losing all the crops. They have stopped growing even maize a staple food crop due to damage caused by monkeys. Unforeseen floods are disrupting travel and other activities. Untimely flowering that is happening in the Himalayan states is a very dangerous trend. The *Rhododendrons*, which grow naturally at an altitude of 2500 meters, are flowering much earlier than they are scheduled to flower. They are to flower in the month of March–April. Now they are flowering in January. Frequent fires are taking a toll on the forests and the total climate is warming up. There is a spread of many epidemics and more allergies and health risks are happening. Among the people who are dependent on the forests and the forest dwelling communities, there is a loss of employment and migration of labor due to lack of work. There is an upward movement of tree lines. Earlier, we used to get, say Deodar at 2000 meters or 1500 meters. Now it won't come at that altitude. You have to move to 2500 meters. Earlier apple crops used to come at 1200 meters altitude, now you have to move up to 1500–2000 meters to grow apples. All these things are happening at the cost of receding glaciers, with those rivers turning non-perennial. This has been happening for quite some time for millions of years. Now it is happening at a much faster rate. The tree fossils found in the Rajasthan desert indicate that at one point of time Deodar trees were present there. Now, it is a desert with Tropical and Arid vegetation. Similarly, the Saraswati River believed to be flowing through Haryana once upon a time is not found now, which is an indication that a lot of climate change is happening and global warming is taking toll on all the natural resources of this universe.

The root cause of climate change is excessive use of energy. We should have a strict audit on the use of energy. Supplying subsidized energy by the States should be dispensed with. Probably, vegetarianism would help in the long term to control climate change, as the production of one kilo of meat requires around 1000 liters of water. The same amount of water can produce about 5 kilos of grains, which will feed at least 20 people. One kilo of meat generates about one kilo of methane, which is again a greenhouse gas and produces around 70–120 kilos of methane per year. Unfortunately, we have one-third of the livestock population of the globe in our country and we are the largest supplier of buffalo meat in the world. We supplied about 1.2 million tons of meat in 2018–2019 alone. All this is coming from our country and we are contributing to methane. This industry takes a heavy toll on the environment. On the domestic front, the consumption of buffalo meat has increased to 13% in rural areas and 30% in urban areas. For short distances, let us use only mechanical means. We have to go for renewable and non-conventional energy sources. In Himachal, there are hundreds of micro, mini and large hydel projects. Despite being a very small state, Himachal Pradesh is supplying power up to Delhi and Haryana states.

ADAPTATION AND FORESTS

Adaptation to climate change is just the process of adaptation to actual or expected climate and try to mitigate adverse effects that are likely. The zero net deforestation policy will definitely help the forest dwelling communities and the people to have their food and water security for the times to come and sustainable forest management practices can limit global warming and increase in the temperatures that are happening.

The State of the Forest Report for the year 2019 in the country shows that India is one of the countries, which is constantly recording an increase in forest cover (Aggarwal, 2020, January 3). It is 24.56% now in 2019 and it has recorded 5188 square kilometers increase. The top states, that recorded an increase in forest cover, are Karnataka, Andhra Pradesh and Kerala. But one unfortunate thing here is that the increase is only 1.14% in the 'very dense' category of forests, which are mainly responsible for absorbing maximum carbon dioxide and the effects of climate change.

In terms of facilitating adaptation, the focus is mainly on reducing the dependency of the village communities on forests and providing them with suitable alternatives to sustain. We are motivating the people not to use wood for cooking and that worked very well. They adopted to use of liquified petroleum gas (LPG) everywhere and you will wonder that till 2010, charcoal was burnt in every part of Himachal Pradesh to heat during the winters, which is now banned and now no charcoal is burnt even if the temperatures goes down to sub zero temperatures also. We asked them not to do grazing without scientific knowledge and that we are practicing everywhere and stopped diverting forests to agriculture. We encouraged them and provided them with pressure cookers, solar cookers, solar water geysers free of cost or at subsidized prices with assistance from

Indo-German Chamber Project or Indo-German Dhauladhar Project or Japan-aided projects, World Bank Projects and UK Projects in the past 30 years. This is how we facilitated the adaptation by the village communities and we have provided them with high-yielding grasses so that the cattle will not go to the pastures and damage them in an uncontrolled manner. Because of all these measures, the forest cover in HP is constantly increasing. That way we are trying to mitigate the climate change aspect. Every Village/Panchayat in Himachal Pradesh has an improved crematoria, which is energy efficient so that people will not consume huge quantities of wood on burning the dead bodies and excessively pollute the atmosphere.

We provided alternate employment to village communities dependent on forests in maintenance of fire lines, construction of forest roads, water-harvesting structures, removal of Lantana bushes, rejuvenating regeneration areas, raising nurseries, fire fighting, cleaning forest floors and raising plantations. Every year 9 million saplings are planted in the state of Himachal Pradesh, where the village communities are deployed for doing this job. We have done plenty of water conservation measures even under MNREGA.[1] Every pond created comprised of 100,000 liters or above. They are allowed exploitation of dead fallen trees, resin tapping, collection of minor forest produce, seeds, pine needles and extraction of Deodar oil according to a systematic 3-year Cycle. Forest dwellers are provided training as eco-tourism guides and cookery for forest rest houses in remote areas. They get free firewood for meeting their rituals because we have plenty of firewood lying on the floor, which is not being exploited now. Also, they can use timber for construction and maintenance of their houses. Using timber for construction is also one way of carbon sequestration.

We give concession to forest dwellers in collection of medicinal plants like Ashwagandha, Sarpagandha, Berberis, Harar, Behra and Amla. They are allowed to collect these in a scientific manner including Guchhi (*Morchella esculenta*), which is a much sought-after Mushroom found only in higher reaches of Himachal and possesses rare medicinal properties, costing approximately Rupees 15,000 per kilo. It is available only for 15 days during the break of winter, when spring starts. They collect it and sell it to the local market and get a good amount of money. So that way

---

[1] Mahatma Gandhi National Rural Employment Guarantee Scheme is an Indian government welfare scheme, which guarantees a 100-day work on demand for rural poor.

their dependence on the forest is reduced. We provide compensation to the communities if they are attacked by wild animals, and even monkeys or snakes, attack their animals. They are provided sufficient compensation so that they will not go for poaching or killing of Wild animals. Animals are also very important to the ecology of forests. Allotment of land has been done to forest dwellers, who are dependent on forests for their livelihood, for the past 3 generations, up to one hectare of land, under the Scheduled Tribe and Traditional Forest Dwellers (Recognition of Forest Rights) Act, 2006. Afforestation can bring the people back to mainstream and we had many successful experiments, where people were earlier resorting to illicit felling and poaching, are now associated succesfully in raising large chunks of plantations, that is why, we use every occasion to plant a tree.

On the banks of River Ravi, stabilization works have been taken up at many locations. Now after ten years, it is a full-grown forest. We have constructed water-harvesting structures with the help of the community, by providing them the wages and they have constructed these and are using the water for their own cultivation of crops. It can store up to 100,000–1,000,000 liters. We have done plantations under the Chamera Project of NHPC in the Chamba district. We conserve even a single tree here, even if it is standing in the middle of a highway. There are plantations of Alnus raised on the banks of River Ravi in Chamba district to stabilize the slopes. There are excellent pastures managed by the communities, which are not interfered with by the forest department at all and are used very carefully by them. There are virgin forests untouched by human beings in Pangi valley of Chamba district. We allow them to raise medicinal plants in their nurseries and private lands, and provide them opportunities to make a livelihood out of the same.

We need to adopt the house construction to save energy and also should go for a large number of small and medium water harvesting structures to store water, increase vegetation and lower temperature. This will change the microclimate of the area. Try to live in harmony with nature. Switch to energy-saving LED Devices. Go for energy-efficient appliances. Avoid the idling of machines waiting for work. Focus on reducing scrap and rework. Paddy cultivation in non-paddy areas should be avoided. Go for dry land agriculture and raise crops like bajra with less water requirement. It's time to cut carbon.

The best way to conserve the environment is not to interfere excessively with the natural resources, without assessing the impact of such interventions.

REFERENCES AND FURTHER READINGS

Aggarwal, M. (2020 January 3). India's forest cover is rising but northeast and tribal areas lose. *Mongabay*. Retrieved on September 10, 2023 from the World Wide Web: https://india.mongabay.com/2020/01/indias-forest-cover-is-rising-but-northeast-and-tribals-lose/

Bhagwat, S. A., Breman, E., Thekaekara, T., Thornton, T. F., & Wills, K. J. (2012, March 5). A Battle lost? Report on two centuries of invasion and management of Latana camara L. in Australia, India and South Africa. *PLoS ONE, 7*(3), e32407. https://doi.org/10.1371/journal.pone.0032407

Mashal, M., & Kumar, H. (2021, February 8). Glacier bursts in India, leaving more than 100 missing in floods. *The New York Times*. Retrieved on September 10, 2023, from the World Wide Web: https://www.nytimes.com/2021/02/07/world/asia/india-glacier-flood-uttarakhand.html

Leveraging Human–Natural World Intersections for Climate Change Adaptation

Priya Tallam

Abstract Working closely with all the species of animals—wild, urban-wild, and domestic—as well as indigenous and coastal communities, has helped VSPCA discover the immense, inter-generational "cultures" of traditional peoples, animals, and plants. Traditional ecological knowledge gathered over generations by indigenous peoples is at the intersection of human society/community and the natural world, revealing tried-and-tested scientific knowledge concerning conservation biology. It is the cultural way of doing science. There have been impediments to integrating the Western disciplines of conservation biology and environmental anthropology even though both are concerned with identifying, describing, and preserving diversity—biological and cultural. Additionally, Western science deems traditional ecological knowledge as "primitive" without its consistent application of the scientific method. However, the disciplines of conservation biology and environmental anthropology are well-positioned to provide greater insights and new perspectives to the world if the knowledge gathered over generations from animal, plant, and people worlds augments these sciences. We argue

P. Tallam (✉)
Visakha Society for the Protection and Care of Animals, VSPCA,
Visakhapatnam, India, Andhra Pradesh
e-mail: priyatallam@vspca.org

© The Author(s), under exclusive license to Springer Nature Singapore Pte Ltd. 2024
N. Bikkina and R. M. R. Turaga (eds.), *Climate Change Adaptation*,
https://doi.org/10.1007/978-981-97-1076-8_10

that biological and cultural extinction threats are imminent and that modern science alone cannot replicate ecosystems' effectiveness. Therefore, finding unique ways of fostering interdisciplinary thinking—sharing demonstrated community-driven conservation workflows and principles with modern scientists, and, applying modern science to indigenous methods—is an effective way forward. This can aid the protection of both biodiversity and culture to reflect the spirit and triumph of life on earth.

Keywords Ecosystem conservation · Biophilic · Sea turtles · Habitat · Human-animal conflict

Ecosystem conservation can be learned by keenly studying animal and indigenous cultures that apply sustainable practices as their lives depend on the respectful utilization of Earth's finite resources. The Visakha Society for the Protection and Care of Animals (henceforth, VSPCA) brings together knowledge systems from the socio-cultural settings of indigenous peoples and animals, augmenting it with modern scientific knowledge, to ensure success in conservation work for the long term. VSPCA, headquartered in Visakhapatnam city in the eastern Indian state of Andhra Pradesh, through action, demonstrates hope to the city's concerned citizens by empowering those interested, to methodically and incrementally address issues in species and habitat loss. All citizens can be naturalists if they join hands in community as VSPCA does. When many community champions join hands to replicate successful ground-up, community-driven infrastructures, the challenges the region is facing, can humbly be grasped and collectively addressed locally, to ripple onto the larger world.

Visakhapatnam city lies along the Bay of Bengal with a strong fishing culture, an amazing variety of flora and fauna, and the potent agrarian economy. It is a biophilic city having to accommodate large numbers of migrant and informal workers coming away from the countryside in search of a "better" life in an increasingly crowded city. Cities may not offer better living conditions for everyone, and one sees slum settlements dotting Visakhapatnam too. Most urban areas are impacted. Visakhapatnam is facing the brunt of climate-related coastal storms, ensuing depressions, inundating floods followed by intense water shortages, and increasing ecosystem degradation. The foreseeable future is already upon the residents of this region.

Insight from the Economics of Biodiversity: The Dasgupta Review

The insightful 103-pages, 2021, Abridged Dasgupta Review (henceforth, referred to as The Review) explains that all capital goods made today in oilfields, mines, plantations, factories, fisheries, or farms, are borne of nature. And, that nature's capital differs from human-produced capital goods in three unique ways: "mobility, invisibility, and silence" (p. 15, Abridged Version). Everything in nature moves—rivers flow, fish swim, trees connect across cities through their hyphae, and the ocean circulates. According to Dasgupta (2021), the ocean circulates invisibly; *it flies in silence*. Building from this idea, in thinking about the Bay of Bengal, the amount of water that evaporates and combines with the transpiration from the Aruku Valley is part of the state's monsoonal system, "flying" silently and profoundly affecting seasons, ultimately, world over. This implies that what is happening in Visakhapatnam (i.e., the consequences of the city's combined human activity affecting its climate) is profound. What is done to the land here affects the climate systems of the larger world.

In the West, single species dominate certain ecosystems: e.g., salmon or trout in the American rivers. In contrast, the Bay of Bengal region is tropically rich, abundant in a multitude of species supplied to the world's food systems via international trade agreements and underhanded commerce operations. Sustaining urban humanity is fast-depleting Earth's natural resources. The Dasgupta Review expresses two phenomena that humans take advantage of in nature's silence: invisibility, and mobility. Dasgupta (2021) speaks to the economics of biodiversity around nature's capital. According to their Review, it ought to be the understanding of economics—the true value of nature, and the ecosystem services all the living obtain from nature, that must inform our institutions producing capital goods on which our lifestyles are based. Since nature's capital is essentially free, and part of the world's commons, there is a massive disconnect about the *value* of nature's capital that goes into making our goods. We err greatly in tagging a price to nature's bounty without truly recognizing the value. Secondly, The Review speaks to the harm humans cause to nature, posing a question to reveal a changing trend: why is there an increasing need for disclosure in the supply chain? This is because, every mother, students of upcoming generations, our youth, seniors, and every concerned citizen has genuine concerns about the harm we collectively are causing as humanity to our once-healthy planet and healthy food systems. There is growing concern about extractive behavior of priviledged humans and large corporations. Each is keen on where to invest efficaciously. Every thoughtful investor, who could be a mother worried about her children's future, is now asking, which corporations are socially and environmentally responsible? Herein lies the political power of an individual.

As a world community, we haven't successfully balanced the economics of biodiversity and the economics of capital goods production. The Review (2021) reveals our true disconnect with nature in our economic, commerce, and legal systems. The Report (2021) prompts rethinking how we ought to reframe our global challenges, such as global warming. Global warming is evidenced by measuring greenhouse gases. Global warming in itself is not the problem; it is the ways by which we misuse land or over-consume and waste products made out of natural capital that are responsible for global warming. We *need* greenhouse gases such as carbon dioxide to warm the planet and keep us alive. If we were able to distinguish root causes that eventually point a finger to our combined behavior, we would understand the profound consequences of not paying heed to nature's invisibility, silence, and mobility. Changing our behavior to align with the needs of the planet can aid our thinking like naturalists. Invisibility around a tree or its soil has a plethora of silent, mobile, and invisible processes that are less obvious to the naked eye—nitrogen fixing, phosphorous management, earthworms fertilizing soil, seeds dispersal, and so on. This is why it is of utmost importance to learn about the *economic value of biodiversity* in terms of *ecosystem services*, when we want to be naturalists or work in the area of conservation.

We have crossed four of the planetary boundaries (Steffen et al., 2015), and the Intergovernmental Science-Policy Platform on Biodiversity and Ecosystem Services (IPBES) Report finds that up to a million plant and animal species will go extinct in decades (Settele et al., 2019). Humanity has crossed, for example, the threshold for silent bio-geo-chemical processes like nitrogen-fixing. Thinking about both concepts in The Review (2021)—(1) the economics of ecosystem services provided by nature, and (2) the harm we directly cause to nature and several species, we cannot pinpoint the harm because of nature's silence and invisibility. We cannot objectify it clearly; and therefore, cannot verify it. This is why, according to The Review (2021), each of us must assume the role of an asset manager, whether farmer, fisher, miner, householder, or corporation, as managing the global portfolio of the world's assets—no matter how small the asset is. When I as a single person litter the beach with plastic, I am basically mismanaging my land asset. Our most precious asset is the natural bio-diverse environment. If we do not empower ourselves as our own asset managers, we can bring ourselves to a massive, collective, global failure of the management of global assets. A disempowered citizen can become self-empowered by acting for planetary health and she can begin seeing the world anew.

VSPCA: Community Cultures and Ground-Up Infrastructures

VSPCA has evolved a hybrid paradigm in a ground-up, "inverse infrastructure" for shore-to-ocean ecosystem conservation. As one of their programs, this provides tremendous scope for marine conservation, where there is a lack of such models, which VSPCA has been expanding through their Olive Ridley Sea Turtle Conservation Program (Visakha Society for Protection & Care of Animals, 2020). VSPCA is working to build coastal resiliency for the fishing villages, urban animals, and the local community. This is worth paying attention to and should matter to conservationists, animal activists, and local people. As understood, it has been hard to bridge the fields of conservation biology and environmental anthropology. Conservation biology involves understanding animal communities: how do primate communities socio-ecologically survive in the urban? what kind of interspecies embodiments a parrot with broken wings displays when captured by a tribal astrologer? or, what is the link between street dogs and sea turtles in coastal cities? Environmental anthropology studies indigenous and tribal cultures with longitudinal information about their geographies. VSPCA brings together knowledge systems from animal and human cultures, to derive solutions demonstrably, to benefit all individuals in the ecosystems, especially those rendered invisible (Tallam et al., 2021) (Fig. 1).

While developing community-led, ground-up, inverse infrastructures, VSPCA pays special attention to people who are often neglected, whose voices are dismissed, and those who are marginalized. In working on species and habitat conservation, VSPCA takes responsibility in building resilient communities to include all residents whether tribesperson, the

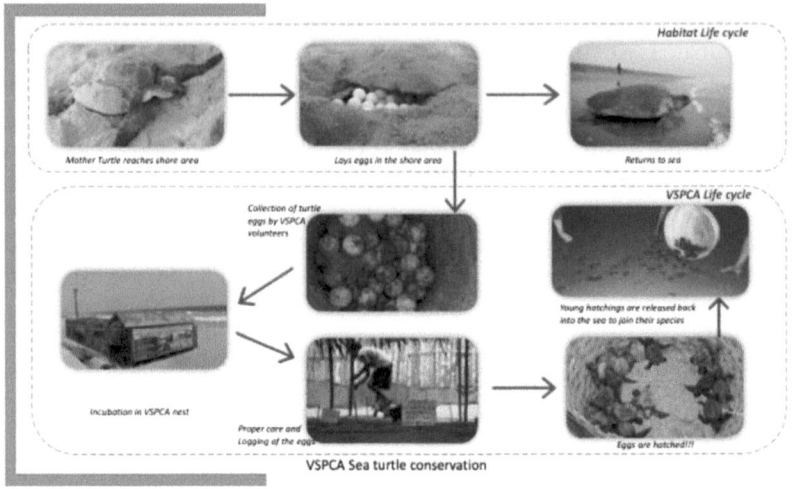

Fig. 1 VSPCA's sea turtle conservation respecting the Olive Ridley life cycle. Illustration by VSPCA staff

unhoused, or physically-challenged. In analyzing their conservation principles, they study the issues around migratory bird species' habitats and/or the estuarine-delta-river–lake systems crucial to rural and coastal farmers in connection to the ocean. The ocean is one pillar in the "inverse infrastructure" that VSPCA works to organize.

VSPCA's three pillars of work foci are: (1) species and habitat and protection, (2) a healthy ocean, and, (3) resilient human communities. Every step demands serious efforts in community participation—of the local citizens, especially residents who are marginalized, NGOs, governments, and leads at corporations that employ large numbers of working people. VSPCA has an integrative and holistic way of thinking about conservation. In attempts to understand VSPCA's vision evolved over half a century by the founder Pradeep Kumar Nath, biodiversity is the underpinning essential for humanity's survival. Nature's cycles of living and dying on the planet is how biodiversity is sustained. And, the crucial factor in the living and dying of species all across this world is the amount of suitable habitat left for them (Wilson, 2019). The renowned American biologist E. O. Wilson implies that our lives hinge on the amount of biodiversity we can retain on this planet. Quoting E. O. Wilson, "I will argue that every scrap of biological diversity is priceless, to be learned and cherished, and never to be surrendered without a struggle".

VSPCA uses novel processes to protect every animal in the city and the surroundings—whether wild, stray, street, passerine, migratory, liminal, urban, domestic, or more. As conservationists, VSPCA chooses to work with communities *emically*. VSPCA staff work daily with native and local communities listening to their wisdom: the fisher people know the ocean intimately, the tides, or the temperature of the sand when a greater population of female turtle hatchlings is expected. This is generationally-tested and passed-down scientific knowledge. Absorbing their inter-generational longitudinal wisdom, VSPCA engages with them and local residents to expand everyone's ecosystem education. VSPCA develops structures from the ground-up rooting their work in awareness, monitoring, and data analysis.

From the vision, their mission breaks down into tenets. The first tenet is built around Public Health. VSPCA's way forward in this is through animal welfare. *Taking care of all the city animals reduces human-animal conflict and reduces the propensity of zoonotic disease spread.* Such kind of animal welfare strategies have gradually directed VSPCA into natural habitat protection for many species. The more we recklessly destroy natural habitat, the more we invite wild animals to forage in our cities—raising the risks of zoonosis.

VSPCA has fought several lawsuits on the mining of sand, all the way to preventing warships marked for "retirement" on beaches, all of which destroy healthy beach habitats. The Legal Model, therefore, is a crucial tool for VSPCA's comprehensive conservation efforts. To wrest back their political power, VSPCA organized a rally against the National Thermal Power Corporation (NTPC) where many "community" or cared-for-street-dogs joined this rally—as though in step with the residents (see Fig. 2). This made habitat conservation a visible matter and a public performance, exposing the polluting impacts of the NTPC. Street dogs matter, and this must matter to all of us because VSPCA solved a key problem to biodiversity conservation via the dogs. These dogs were big predators of sea turtle eggs. Because VSPCA looks after the 150,000+ dogs in the city, now well-fed, they no longer predate the sea turtle eggs. The dogs have evolved to aid VSPCA and the fisher people in monitoring predators like crows and raptors. They bark when the mother turtles come ashore after midnight. They alert fishermen to injured sea turtles. They watch the mother turtles lay eggs while holding back curious humans. These are amazing spectacles to watch when one patiently follows the dogs, VSPCA staff, and fisher communities in their diligence around sea turtle conservation. Working with animals has solved a key predator problem. Similarly, each animal in the city ecosystem from a cow to a parrot to a cobra, has a deeper connection to the city's ecosystem services, and each species matters deeply. In community-oriented field work, these issues slowly come to light. One cannot research their way toward such insights.

Today, VSPCA's-cared-for-dogs are not only part of their Sea Turtle Protection Force (STPF), but are also guard dogs for Vizag businesses. Animals like urban street dogs have learnt to live effectively with humans. Instead of seeing dogs as nuisance or vermin, caring for them strongly evokes a growing compassion for all biodiversity. Citizens can start to understand the value of biodiversity by caring for domesticated animals seeking to survive in cities.

Fig. 2 "Community Dogs" participate in a rally that exposed a major polluter, paving the way for sea turtle conservation (PC: VSPCA Staff)

Engagement with the Fishing Community

When VSPCA engages with fisher communities, they explain the legal model in their language where laws and regulations are broken down simply and clearly. This raises the fisher people's awareness of regulations around harmful gear and endangerment of marine species. VSPCA staff is able to talk through the dangers of specific fishing gear for endangered species, and the usefulness of additional gear such as Turtle Excluder Devices (TED) to free turtles caught in by-catch. Addressing a myriad of issues to bring about a holistic solution, raises the staff and community-based awareness up many notches, increasing the quality of the fieldwork. Conservation is not about the species alone; it is also about self-reflection and a rethink of our own actions and methods. The mutual trust built through such interactions has improved the willingness of the fishermen to understand the importance of say, the IUCN Red List. VSPCA is gradually streamlining the best indigenous practices by paying respect to fisher's traditional indigenous wisdom. As naturalists applying the modern scientific method, VSPCA stands to benefit with greater insight, applying indigenous wisdom to science. VSPCA's employement of fishermen to construct in-situ and ex-situ sea turtle hatcheries, and patrol the beaches for gravid turtles coming ashore, has provided alternate livelihoods to several fishermen. This has forged understanding between VSPCA and fisher communities, in that, *their* interests matter to this region and to the larger economy.

Cleaning up beach habitats, preventing city encroachments on the beach, and working towards a healthier food web in marine and ocean habitats have brought VSPCA to the nexus between disaster management and biodiversity conservation. The fisher people have sustainable ways by which to safeguard their land and ocean habitat for disaster preparedness. This is the next step in VSPCA's conservation infrastructure—to bring fisher peoples, scientists, and governments together to build upon the current solutions, conducive to disaster management. Coastal storms that lash this region are woven into the lives of the fisher people. This is an area in which modern science and the indigenous wisdom of generations can come together and prove useful and practical in disaster prevention and management (Hadlos et al., 2022).

India and Beyond

India is facing the consequences of steady socio-economic growth and development. As an economy, India is rising. However climate change and intense urbanization are big threats. Almost 68% of the population is going to be urban, in the world cities, by 2030. Humanity's future is indeed urban. We must look at solutions to climate change and conservation from *within* the city. Thus far, there have been many conservation efforts but there is less evidence of these having reached a multiplier effect or economies of scale. Most conservation efforts are backed by large international contributions, or are complex, cost-prohibitive scientific interventions, and therefore, have not had long-term success. These factors importantly have lost the trust *and* buy-in of local participants such as the local fisher communities as an example. In the long term, every sustainable program requires the trust and participation of the locals as they have the greatest stake in the region's (and their community's) resilience. Problems like pollution, which impact conservation efforts, arise from chemical spills or sewage flows into river systems and the ocean. Fisher people suffer the brunt of the harmful consequences. Conservation efforts which involve these communities to have a strong say in the solution, can reap long-term sustenance, adaptiveness, and flexibility. Local people have the know-how to figure out the consistent change management needed to sustain the "inverse infrastructures."

As concerned citizens, we are constantly evaluating our choices even if we can't fully grasp the phenomenon of climate change. As investors, we are evaluating where to invest our funds. According to Dasgupta (2021), all of this makes us our own judge and jury of our actions. Even if we can't personally act as conservationists, it is our duty to support organizations such as VSPCA or your local council involved in environmental and social justice matters.

According to Dasgupta (2021), one-sixth of the carbon footprint of the average diet in the European Union can be linked to deforestation in tropical countries. Also, they explain that 70% of bird species on earth today are poultry. In sharp contrast is the ancient Hindu belief of *Vasudhaiva Kutumbakam* which sees all the living species of the world as one family. Thinking of animals as "poultry" versus living beings with an interest in living, raises a moral question to be answered by those wishing to pursue conservation. In thinking about biodiversity, arbitrarily picking animals to eat, love, conserve or experiment upon, cannot bear success in terms of conservation or species welfare.

Biodiversity conservation from a community-driven empowered manner, builds upon and into the environmental and social justice movements. One can begin to eliminate the hierarchical structures of caste, social class, religion, and crucially, begin looking beyond human form (to the animal). Without the separation of urban from the natural in our thoughts and actions, a healthier ecosystem and all its services can be available to residents in cities.

From twenty-six (26) baby turtles that hatched in 1997, by 2020, VSPCA was safely sending about 70,000 hatchlings to the sea each year (see Fig. 3 for an illustration of baby turtles on Visakhapatnam beach). This program has brought over half a million hatchlings safely to sea from 2000 to 2020, which is helping increase the fish stock, which is helping the fisher community change the way they understand conservation. They are coming into alternative employment because they are managing this program—from the construction of the hatcheries to the care and release of baby turtles.

The sea turtle conservation effort lays the groundwork to protect migratory birds in the adjacent wetlands near the ocean, while bringing VSPCA to look into endangered shark species' conservation with fishing communities. This involves members of the Visakhapatnam Port Trust, The Central Marine Fisheries Research Institute, Fishing Mandals, and many NGOs affiliated with the UNDP. One program can provide insights

Fig. 3 Turtle hatchlings making their way to the Bay of Bengal (PC: VSPCA, Virendranath)

to humbly expand efforts toward similar challenges that may seem unsurmountable if studied in silos.

We are not above creation.

We must align our lifestyles to reflect the needs of our finite planet and join forces with those who have evolved best practices through traditional wisdom and modern science.

REFERENCES

Dasgupta, P. (2021). *The economics of biodiversity: The Dasgupta review.* HM Treasury.

Hadlos, A., Opdyke, A., & Hadigheh, S. A. (2022). Where does local and indigenous knowledge in disaster risk reduction go from here? A systematic literature review. *International Journal of Disaster Risk Reduction, 79,* 103160.

Settele, J., Diaz, S., & Ngo, H. T. (2019). *Global assessment report of the Intergovernmental Science-Policy Platform on Biodiversity and Ecosystem Services.* IPBES Secretariat.

Steffen, W., et al. (2015). Planetary boundaries: Guiding human development on a changing planet. *Science, 347*(6223), 1259855. https://doi.org/10.1126/science.1259855

Tallam, P., Nath, P. K., Tallam, K., Logan, A., & Veeravalli, S. (2021). VSPCA sea turtle conservation. *Biophilic Cities Journal, 4*(1), 16–25.

Visakha Society for Protection and Care of Animals. (2020). The Olive Ridley sea turtle community-based protection program: Urban coast of Visakhapatnam 2019–2020. *vspca.org.* Retrieved on September 18, 2023 from the World Wide Web: https://vspca.org/wp-content/uploads/2015/12/The-Sea-Turtle-2020.pdf

Wilson, E. O. (2019). Discover half-earth. *The Half Earth Project.* Retrieved on September 25, 2023, from the World Wide Web: https://www.half-earthproject.org/discover-half-earth/

FURTHER READINGS

Kundu, S. K., & Santhanam, H. (2023). How to fish—Combining indigenous, traditional and scientific fishery advisories to preserve marine provisioning ecosystem services of Odisha, India. *Social Science Research Network.* Retrieved on September 25, 2023, from the World Wide Web: https://papers.ssrn.com/sol3/papers.cfm?abstract_id=4330063

Lepofsky, D., & Caldwell, M. (2013). Indigenous marine resource management on the Northwest Coast of North America. *Ecological Processes, 2*(1), 1–12.

Nath, P. K., Tallam, P., & Virendranath, (2020). VSPCA Sea Turtle Conservation Report. *Visakha society for the protection and care of animals*. Retrieved on September 25, 2023, from the World Wide Web: https://vspca.org/wp-con tent/uploads/2015/12/The-Sea-Turtle-2020.pdf

Nath, P. K., Varma, H., & Tallam, P (2022). VSPCA's annual report. *Visakha Society for the Protection and Care of Animals*. Retrieved on September 25, 2023, from the World Wide Web: https://vspca.org/wp-content/uploads/2015/12/Annual-Report-2021-2022.pdf

Nath, P. K., Varma, H., Virendranath, & Tallam, P (2022). VSPCA's annual report. *Visakha Society for the Protection and Care of Animals*. Retrieved on September 25, 2023, from the World Wide Web: https://vspca.org/wp-con tent/uploads/2015/12/Edited-Final-AR-20-21-1.pdf

Thornton, T. F., & Scheer, A. M. (2012). Collaborative engagement of local and traditional knowledge and science in marine environments: A review. *Ecology and Society, 17*(3), 8. https://doi.org/10.5751/ES-04714-170308

Listening to Locals

Abstract This transcript of a talk describes what happened when residents from a vulnerable low-lying coastal area worked together with local and national agencies to confront the risks and possibilities of climate change, after first reaching out for advice from overseas experts. During the last two decades, their joint work has led to Europe's largest open coast realignment, the establishment of a multi-stakeholder climate change partnership and one of the UK's earliest Integrated Coastal Zone Management policies.

Keywords Stakeholders · Peninsula · Climate change · Local authorities · Planners

What I want to show you is how locals can work with different stakeholders such as local authorities and national authorities to help everybody gain more understanding of the impact of climate change and to collectively produce integrated long-term solutions. I am going to talk

C. Cobbold (✉)
University of Cambridge, Cambridge, UK
e-mail: cacobbold@gmail.com

about my work on the Manhood Peninsula, which is a very low-lying peninsula right at the bottom of the UK, about 50 miles south of London. Much of the peninsula is about 5 meters or less above sea level. Partly because of that, we are blessed with a lot of biodiverse wetlands, but we also have the issue of having to protect housing and communities in an area that will become increasingly vulnerable to climate change. We are one of the most vulnerable coastlines in the UK, partly because the bottom of the UK is tilting down while the top, Scotland, is rising. So more than climate change and rising sea levels, we have also got that shift in the geology (Fig. 1).

The Manhood Peninsula has about 16 settlements of which the largest is Selsey and the smallest is just 100 or so people. It is relatively rural. Nearby is Portsmouth, a large urban, historically military city. It has got hundreds of thousands of people together with its neighboring community of Gosport, which again is heavily urbanized and they are facing big decisions as to how to manage sea level rise, which will entail billions being spent on sea walls. So we have a different issue to confront, which is we don't have the population that would justify that sort of expense, but at the same time, we still have demand for housing to be built here without necessarily being able to afford those big sea walls. As a resident, I could see that we were facing a long-term problem we had to address, although most of my fellow residents obviously were not so aware of the climate change issues. I had a friend at that time, who lived here, who later went back to the Netherlands, who was a Dutch spatial planner by training. Obviously, much of the land in the Netherlands is actually below sea level. So she also was acutely aware of the issues of managing water and managing development. In 1997 we decided that we would approach the local authorities to see what the long-term plans of the area were. We approached the district, which is the local urban planning council, the county council, which is like the small state authority and the environment agency, which is the UK national body that looks at flood risk management and the environment. We realized quite quickly from speaking to them that they were not particularly operating in a very integrated and long-term manner and they certainly weren't addressing climate change. This was in the late 1990s. We decided then to speak to the public and the local residents in the area to find out their views and how aware they were of the issues. We persuaded my parents and our kids to make tea and cakes to lure these residents in to come and talk to us in meetings held in village halls. It became increasingly clear that the level of understanding

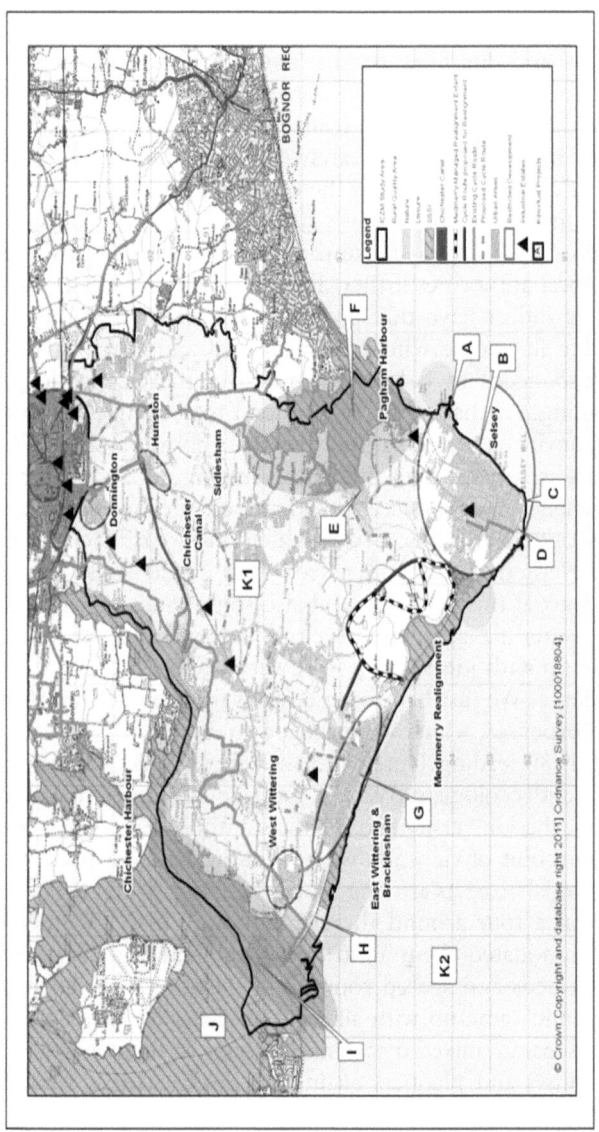

Fig. 1 Map showing the study area and regions of interest on the Manhood Peninsula (*Source* Chichester Coastal Change Pathfinder Project [2011]. *Towards ICZM on the Manhood Peninsula*. Chichester: Author)

among the public was also very low. At that point, we went back to the local authorities and asked what they were going to do about climate change.

After meeting and speaking to the environmental director of the district council for about an hour, he turned to us and said that it was very entertaining, at which point we realized that he wasn't looking at us as two professional people, one a researcher with scientific training and an understanding of climate change and the other a qualified spatial planner. He was viewing us as two housewives who lived on the peninsula. We thought we were not going to make the sort of traction we wanted to make with the local authorities if they didn't see us as having any knowledge. Even if we didn't have that professional background, there is an innate knowledge in people who live in an area and that is an understanding that also needs to be tapped into. But that acknowledgment wasn't forthcoming. At that point, we decided that we needed to bring in 'experts' to convince the local authorities that this was an issue. Renee, my friend, who was a member of the Dutch Institute of Spatial Planning (NIROV) approached Nirov, aware that every year they would organize a brainstorming workshop for members in which they would tackle a particular issue in the Netherlands. We discussed if they would like to come over to the UK and look at this peninsula and also some issues going on about climate change, water management, development pressures, etc., do a case study and come up with ideas. They were very keen and they offered to come free of charge. We just had to raise funds for their accommodation and their travel expenses, which we managed to do.

We brought in 30 leading Dutch and British planners, water management engineers and ecologists. We got local communities, local village councils, schools, the University and the local authorities to write briefing papers from their point of view and we put that together as packs for the Dutch and the British experts arriving for the meeting. When they arrived, we took them on a tour around the area and explained all the various issues. We accommodated them all in one venue and lots of them were sleeping in bunk beds and shared rooms and they literally brainstormed for a week solid and came up with all sorts of ideas. We deliberately put them into three separate mixed disciplinary groups so that we would have engineers, ecologists and planners all in one group so that they could come at it from each other's point of view. At the end of each day, they would come up with all their ideas—what they discussed during that day and their drawings and things—and present them to the local community. Then the local community would ask them questions or point out issues that they wanted to point out and then they go back literally to the drawing board for the next day. At the end of it all, we had a big meeting in the town hall in which they gave their presentations. It was interesting because the three groups could come up with any plans they wanted for the future, the next hundred years for the peninsula.

Initially, all three groups, to the surprise of the locals, said that we should let the sea in where we could. Where we have rural land behind the coast, the experts said that we should not keep fighting the sea because we will lose. That did come as a bit of a shock to the locals, particularly with the reputation of the Dutch holding back the sea with massive dykes. We decided to tell one of the groups to look at an alternative option to demonstrate to locals what not letting the sea in meant (ie much higher sea walls, no beach by the end of it, because you have lost your beach and the inability to actually see the sea because the walls would get so high) so that people understood both outcomes. By the end of the week, the experts came up with lots of really interesting ideas, but the main three outcomes included the creation of a managed realignment scheme at Medmerry, which is the area between the two coastal settlements. They said that would be a much more effective way of producing a sustainable sea defense and actually increasing the environment and producing a habitat that would draw more tourists in so that it would help the economy as well. They also said that we should develop an Integrated Coastal Zone Management Strategy for the area. So, instead of just looking at how you are going to defend the coast, you need to look at what you want to do behind the coast, as far inland as is relevant. So you have got an integrated approach to your coastal management. It is not just the coastline. They also talked about the creation of a form of partnership between all stakeholders in the area. That is from the community up—the community, local businesses, local landowners right the way through to the local authorities and the national authorities. This was in 2001.

Six years later, the Environment Agency said it was going to produce a managed coastal realignment scheme at Medmerry along the lines of what the Dutch had suggested. Again there was a degree of reluctance, although by that time some residents had realized that actually this could offer opportunities for the area and could form a more sustainable sea defense, but there were a lot of residents who still were unaware of climate change and unaware of the suggestions that the workshop members had come up with in the past and were very reticent to see the loss of what had been a shingle bank keeping the sea out for hundreds of years. Why would you let the sea in? So at that point, we again brought some of the Dutch back. Some of them were the ones who have been at the original workshop; some were new, to stress test effectively the Environment Agency's scheme in front of the public—to say that 'this is what we think of the scheme' and again show the public what the alternative would be in the long term. Before breaking open the shingle bank the Environmental Agency contractors built a big clay bund further inland. The sea has gradually come in. What this has done is made a much softer and more sustainable coastal defense because the energy of the sea is being absorbed by this land rather than hitting the big shingle bank. Prior to this scheme, the Environment Agency literally had to have diggers right along the bank in the winter months to keep shoring up the shingle bank at a considerable cost. So this is a more sustainable sea defense. It also created 183 hectares of new wetland and that compensatory habitat actually helped the docks and ports at Southampton, which is a big container ship port to the west of Portsmouth. It allowed the docks to expand onto the wetland in Southampton by creating a compensatory wetland on our peninsula. It was a win–win situation because it helped the economy of Southampton by extending the docks, but it also has helped our economy by boosting tourism because green and sustainable tourism and rural coastal tourism is one of our area's main economies.

Also, immediately after the Dutch left, we set up the Manhood Peninsula Partnership,[1] which allowed us to confront some of the issues we were facing, but also opened opportunities that mitigating climate change might address. The members of the partnership ranged from the community at the grassroots through to the local authorities, stakeholders, RSPB, and a volunteer group of locals who maintained the ditches and the ponds and both for the wildlife and for better drainage.

We have lots of gains. The drainage of the area through the creation of more wetlands has improved the wildlife of the area. We created vision processes where local people look at their streetscapes and how they can improve them. We have also got a project going, which is looking at the lobster and crab habitats at sea to understand why they are changing because that affects the fishermen's livelihoods. We had

---

[1] https://peninsulapartnership.org.uk/.

a wonderful oral history and modern celebration called Sea's The Day with the entire fishing community. We have been appointed as a Nature Recovery Network Delivery Partner by the Government in their new scheme to try and create more natural areas and networks of natural areas. We have sub-groups such as GLAM, standing for Green Links Across the Manhood, creating sustainable travel networks across the area. We introduced a destination management plan to encourage green, sustainable tourism. This in itself actually brought new private investment in this area, because recognizing the peninsula as a tourist destination has encouraged businesses to open new restaurants, surf shops, bike shops and to take advantage of the Medmerry Scheme.

There were lots of intangible gains—increase in public optimism, local pride, putting the area on the map and making people realize that climate change introduces opportunities as well as negative things. Local residents can stimulate great change. Climate change mitigation can benefit an area environmentally, socially and economically as well as produce greater security. People from different disciplines and backgrounds working together, outside of existing political, disciplinary and administrative constraints, can result in innovative and beneficial solutions. An integrated approach helps to identify short and long-term decisions and a range of problems and solutions. Bringing together all parties, residents, businesses, professionals, administrators and landowners, early in the process increases the chance of reaching a consensus. A vision for the area is vital to ensure long-term gains are achieved as well as short-term wins and to create more opportunities. A holistic and integrated approach often solves more than one problem with one solution, producing more financial benefits. Working together helps to identify which issues should be solved within the collective domain and which by local and private initiatives.

FURTHER READINGS

Few, R., Brown, K., & Tompkins, E. L. (2007). Public participation and climate change adaptation: Avoiding the illusion of inclusion. *Climate Policy, 7*(1), 46–59.

Glucker, A. N., Driessen, P. P., Kolhoff, A., & Runhaar, H. A. (2013). Public participation in environmental impact assessment: Why, who and how? *Environmental Impact Assessment Review, 43*, 104–111.

Kettle, N. P., Dow, K., Tuler, S., Webler, T., Whitehead, J., & Miller, K. M. (2014). Integrating scientific and local knowledge to inform risk-based management approaches for climate adaptation. *Climate Risk Management,* *4,* 17–31.

Lieberknecht, K. (2022). Community-centered climate planning: Using local knowledge and communication frames to catalyze climate planning in Texas. *Journal of the American Planning Association, 88*(1), 97–112.

Meijerink, S., Stiller, S., Keskitalo, E. C. H., Scholten, P., Smits, R., & van Lamoen, F. (2015). The role of leadership in regional climate change adaptation: A comparison of adaptation practices initiated by governmental and non-governmental actors. *Journal of Water and Climate Change, 6*(1), 25–37.

Scholten, P., Keskitalo, E. C. H., & Meijerink, S. (2015). Bottom up initiatives toward climate change adaptation in cases in the Netherlands and the UK: A complexity leadership perspective. *Environment and Planning C: Politics and Space, 33*(5), 1024–1038.

Climate Change Adaptation: A Thematic Analysis of Narratives of Fisher Folk and Tribal Farmers

Satya Kishan Kumar Namala and Nalini Bikkina⦿

Abstract The escalating global warming of the planet is aggravating the consequences of climate change, particularly among the marginalized sections of the population. The planning of climate change adaptation policies often subverts locally immersed practitioners' perspectives. The research on indigenous communities' capabilities to make adjustments to climate change, while growing, has been still relatively sparse. The gaps are particularly discernible with reference to acknowledging the social element in the context of adaptation and in precluding knowledge beyond academic parameters from the debate. The proposed research intends to draw from the narratives of long-form conversations with two fisher folk and three tribal farmers in the Visakhapatnam coastal region, from the point of view of the stages of public adaptation outlined by Risbey et al. (Mitigation and Adaptation Strategies for Global Change 4:137–165, 1999). This study uses qualitative analysis along with the elements

S. K. K. Namala · N. Bikkina (✉)
GITAM Deemed-to-be University, Visakhapatnam, Andhra Pradesh, India
e-mail: nbikkina@gitam.edu

S. K. K. Namala
e-mail: snamala2@gitam.in

117

N. Bikkina and R. M. R. Turaga (eds.), *Climate Change Adaptation*,
https://doi.org/10.1007/978-981-97-1076-8_12

of the grounded theory approach to derive themes from the interviews of fishermen regarding climate change concerns and adaptation practices.

Keywords Fisher folk · Adaptation · Climate change · Signal detection · Local communities

The escalating global warming of the planet is aggravating the impacts of climate change particularly among the marginalized sections of the population dwelling in poverty (IPCC, 2014). Adaptation as defined by the Intergovernmental Panel on Climate Change as cited in Ayers and Dodman (2010) "describes the adjustment in natural or human systems in response to actual or expected climatic stimuli or their effects, which moderates harm or exploits beneficial opportunities." The term adaptation is now more extensively used to delineate endeavors to enable sections of communities in peril to manage vulnerability in the contexts of stress created by adverse changes in climate (Agarwal, 2009; Lemos et al., 2007; Orlove, 2009). As against mitigation, adaptation occurs from global to local scales, catering to scale-specific concerns and drawing from the capacities and resources available to the specific cohort of actors (Adger, 2001).

Taxonomies on Adaptation

Several taxonomies on adaptation have emerged through research (Feenstra et al., 1991). Klein (1998, 2003) differentiates proactive adaptation from reactive adaptation and also distinguishes between public adaptation and private adaptation. The timing of adaptation determines whether it is proactive or reactive with reference to whether adaptation is driven by a prediction of a climate event at an indeterminate point in time to come or by the outbreak of such an event. Private versus public adaptation is based on the agents and thereby tries to determine who are the actors involved in adaptation. Adaptation can also be incremental or transformative (Pelling et al., 2015). Incremental adaptation works on maintaining the existing systems, while making minor adaptive modifications. Transformative adaptation on the other hand attempts to modify the basic attributes of the existing systems as a response to the actual or anticipated effects of change (IPCC, 2012).

Stages of Public Adaptation

Risbey et al. (1999) outline four stages of the public adaptation process. These include (i) the signal detection stage where what needs to be adapted is chosen along with the identification of what needs to be ignored; (ii) the evaluation stage in which the climate change signal is interpreted along with an evaluation of anticipated consequences; (iii) the stage of decision and response involves an overt modification in the performance of the systemic entity; and (iv) the feedback stage monitors the offshoots of the decisions.

Climate Change Adaptation and the Local Communities

Connecting the vulnerability of communities and systems to the impacts of climate change to determine adaptation measures is a complex socio-political process. Therefore, outcomes of climate change adaptation are dependent on the participants in decision-making and the framing, justification and operationalization of the processes (Morchain, 2018).

The framing of climate change adaptation policies often subverts locally immersed practitioners' perspectives. While there is some research on local communities' capabilities to adapt to climate change (Cruikshank, 2001; Maynard, 1998; Nuttall et al., 2005), there are significant gaps in our understanding. The gaps are discernible with particular reference to acknowledging the social element in the context of adaptation and in precluding knowledge beyond academic parameters from the debate (Morchain, 2018). However, there is an increasing acknowledgment of the urgent necessity to embed indigenous observations into climate change research among members of the scientific community and to connect research and adaptation requirements in the field.

These local communities have been admitted to being ingenuous observers of the change phenomenon and their interdependence on living and nonliving ecosystems (Fox, 2002; McNeeley, 2009; McNeeley & Shulski, 2011). It is thereby imperative to critically explore the diverse climate change challenges for indigenous communities and the structural roadblocks that restrict the capacity of especially marginalized people to adapt to climate change, more so in times of resource insufficiency (McNeeley, 2017). Climate change adaptation that is sustainable transpires when strategic community actions are initiated to reduce the adverse impacts on the critical flow of natural resources for the current times and the future (McNeeley, 2012). Adaptation therefore needs to be consultative with reference to mainstreaming marginalized knowledge (Morchain, 2018). Moreover, framing adaptation priorities on the basis of technical solutions leads ironically to the distrust of the targeted sections of the vulnerable population (Otto-Banaszak et al., 2011) and effectively prevents them from informing the holistic picture of adaptation (Morchain, 2018), rendering adaptation efforts inefficient (Nagoda, 2015). Empowering local communities through self-governance to control resources including land and water and to draw up climate change adaptation initiatives is of critical significance (Eriksen & Brown, 2011; Eriksen & O'Brien, 2007; McNeeley, 2017). However, it is also critical to acknowledge the pros and cons of traditional knowledge and scientific understanding to arrive at a meaningful hybridization (Lebel, 2013).

OBJECTIVES AND METHODOLOGY

The objectives of the proposed research are to scientifically document adaptation initiatives of the local communities in the relatively marginalized terrains and sections of the population in rural India from a bottom-up perspective rather than a policy-driven viewpoint. In the medium to long term, this research will continue to identify similar instances and document them in the form of detailed case studies. Factoring in the National Action Plan on Climate Change and the State Action Plans by the respective state governments in India, this research proposes to inform policy with special reference to a consultation framework that includes local knowledge and emphasizes collective action at the community level with the state catalyzing community action following a bottom-up approach. This helps make adaptation efforts efficient, equal and just.

This research draws from the narratives of three coastal fishermen and two tribal farmers from the point of view of Risbey et al. (1999) stages of the public adaptation process. The researchers used open form key informant interviews. The researchers went into this inquiry without a priori commitments, which allowed for built-in flexibility and to factor in the iterative complexity of the research subject. This research involved a thematic analysis of the climate change narratives elicited on the basis of an interview protocol, by using the method of constant comparisons (Glaser, 1965; Hallberg, 2009). The respondents were approached after obtaining consent and were requested to share their experiences related to their perceptions of climate change and its manifestations, natural calamities or human interventions which triggered the comprehension of climate change, the impact of these changes on their lives and livelihood and foreseeable solutions to the issues that they are facing in this context. The issues related to climate change included extreme climatic events and consequent disasters, rising temperatures and loss in species quality among other aspects related to climate change. The generalizations from this study therefore are qualified, conditional and situated.

SIGNAL DETECTION, EVALUATION AND RESPONSE

A thematic analysis of the narratives recorded as part of the in-depth interview, using Braun and Clarke's (2006) six steps revealed that these communities did perceive a change in climate and have evaluated that climate change had and continues to impact them. In coding the data, we have identified responses related to acknowledgment of change in climatic conditions as 'signal detection'; responses identified as a comprehension of the impact of climate change on their lives and livelihoods as 'evaluation'; responses related to ideation and implementation of attempts to cope with the impact of climate change as 'decision and response'; and responses related to evaluation of these decisions and attempts at course correction as 'feedback' stages. Farming and fishing have been occupations that have been passed on to these participants from several generations. The interviewed fisher folk and tribal farmers reported that in considering the impacts of climate change, they have arrived at certain responses and are attempting to adapt to the scenario to overcome the adverse impacts. Thereby, we can deduce that these communities went through the stages of Signal Detection, Evaluation and Decision and Response stages of public adaptation (Risbey et al., 1999; Tonmoy et al., 2019). However, the themes, as shown in Table 1, do not report these communities reaching the fourth stage of public adaptation, namely feedback, which monitors the offshoot of these decisions and responses. Nevertheless, these communities were able to identify several constraints hindering the adaptation process.

Table 1   Codes, themes and stages of adaptation

S. No.	Codes	Observations	Themes	Stages of adaptation
1	Transition from traditional to modern crops/single to customized net	TF, FF	Adaptation	Decision & Response
2	Unavailability of seeds (modern crops)	TF	Constraint	Evaluation
3	Change in rainfall and temperature patterns	TF	Change	Signal Detection
4	Water shortage	TF	Change	Signal Detection
5	Taking turns due to shortage of water	TF	Adaptation	Decision & Response
6	Decrease in cultivable land due to less water	TF	Change	Signal Detection
7	Sense of community	TF, FF	Adaptation	Decision & Response
8	Change in plant maturity pattern due to weather	TF	Change	Signal Detection
9	Falling prices due to increased supply	TF	Constraint	Evaluation
10	Agriculture is not profitable	TF	Change	Evaluation
11	Unable to give up since it is the only skill	TF	Constraint	Evaluation
12	Loss of soil fertility because of repeated cultivation	TF	Constraint	Evaluation
13	Crop rotation	TF	Adaptation	Decision & Response
14	Deep sea fishing	FF	Adaptation	Decision & Response
15	More cyclones	FF	Change	Signal Detection
16	Government cyclone alert system	FF	Adaptation	Decision & Response
17	More nets less fish	FF	Constraint	Evaluation
18	Fish no longer available in older location	FF	Change	Signal Detection
19	Beach is different now	FF	Change	Signal Detection
20	Rise in water level	FF	Change	Signal Detection
21	Decrease in shore	FF	Change	Signal Detection

(continued)

Table 1   (continued)

S. No.	Codes	Observations	Themes	Stages of adaptation
22	Area constraint—parking boats, nets, etc.	FF	Change	Signal Detection
23	Olden days catch was fixed/and high chance of getting fish	FF	Change	Signal Detection

Perception of Change

The communities perceived climate change even when they could not articulate using the exact phrase. Changes were noticed due to natural processes as an offshoot of climate change in referring to the destruction of habitats which acted as natural protection systems against adverse weather conditions, migration of fauna, surge in sea level, increased frequency and intensity of severe weather events like the El Niño and changing patterns of precipitation and rise in ocean temperature. One of the fishermen interviewed observes:

> As far as I know, cyclones have increased in recent times. Also, the nature of the sea has changed over the years. The waves were more during those times.

The increase in sea level along with erosion of beaches was an obvious observation by the coastal fishermen. One fisherman reports:

> Erosion of beaches has become a hassle for us, making it difficult for us to drag our boats into the sea. Sometimes we tie the boat in the evening and return in the morning to find them overturned. It is quite a task to put them back in order.

Tribal farmers did notice the changing patterns of rainfall with reference to timeliness. They reported facing unprecedented water shortages, particularly during the summer months. Also, they notice an increase in temperatures overall and a rise in night temperatures specifically. Weather has also turned unpredictable of late and has created challenges for these communities. For instance, one of the tribal farmers reports:

Cauliflower blossoms phase wise when there is sunshine. However, if the weather is cloudy, it blossoms all at once. If it blossoms all at once, we cannot market it. Prices fall in that case.

Fishermen and tribal farmers who were part of this occupation for several decades could discern the change while drawing from their knowledge of history, cultural and local politics. The communities also evaluated and reported a volatility and decline in their income from fishing and farming and a fall in their profit margins as a result.

Reported Scarcities

The fishermen reported a current scarcity of fish in the vicinity of the waters where they would cast their nets previously. Consequently, there were frequent instances of low catch hitting their livelihood and sustenance. One of the fishermen remarked:

> Earlier, there was good catch even in shallow waters. Now we have to sail far away and still don't return with a good catch. During earlier times, fish were available even close to the shore. Today people use too many nets to catch fish, hence driving them further away from the shore.

The size of the fish too has significantly decreased, impacting their market and incomes. The farmers face the adversity of water shortage, especially in the summer months. For the farmers, water shortages due to erratic rainfall and lack of irrigation including lack of recharge of underground aquifers have erected huge challenges in the path of viable farming. One of the farmers declares:

> We have cut down the area of cropping from 200 acres. We are farming only in the surrounding areas, where there is water availability.

Attempts at Adaptation

Fishing and farming communities seem to initiate behavioral adaptation to climate change, albeit an incremental one. Fishermen tend to migrate to landscapes that provide access to deep sea fishing which allows relatively better catch quality and quantity. Fishermen also spend longer time than they were doing previously to catch quantities of fish similar to their catch in the past. Fishermen additionally use technology on their mobile phones to receive alerts on impending inclement weather and tropical cyclones in a context where climate change has led to increasing frequency of these extreme weather events. One of the fishermen comments:

> Cyclones have increased in recent years. However, the government alerts the fishermen on any approaching cyclones following which we refrain from venturing into the sea. This ensures our safety.

Farmers have stated that they have altered their choice of crops cultivated in an attempt to adapt to climate change, with particular reference to water shortages owing to a decrease in rainfall, warmer summer temperatures and consequent degradation in soil condition. One of the interviewed farmers stated:

> We used to cultivate coarse grains at that time. Now, we cultivate cabbage, cauliflower, broccoli, red cabbage and lettuce. During the rainy season we cultivate crops that are not water intensive.

In an attempt to overcome the constraints thrown at them by water shortages, the tribal farmers take turns to irrigate their fields. One of the farmers shared:

> We try to take turns to cope with water shortages. If everyone irrigates at once, there is an acute shortage. Each farmer irrigates 20 to 30 cents. If there is no water, the rest of the land is wasted.

While the shifts in crop patterns helped in adaptation to climate change marginally, farmers also face the constraint of soil degradation owing to these shifts. The farmers attempt to overcome this challenge by performing crop rotation. One of the tribal farmers claims:

> We are doing some crop rotation - cauliflower, then cabbage, then cucumber, then bottle gourd and bitter gourd.

Reported Constraints

Fishing and farming communities are highly vulnerable to climate change, a fact acknowledged by these communities in no indefinite terms. Their attempts at adapting to the vulnerabilities thrown at them by climate change by deep sea fishing are also thwarted by mechanized trawlers which can take away a larger catch and go deeper into the sea. These trawlers therefore overfish thus leaving the traditional fishing communities in the lurch. One of the fishermen complains:

> In the past the availability of fish was higher but the nets were small and limited. Today we have less fish but more sophisticated and expensive nets.

Among the farmers, the lack of water for irrigation has challenged their adaptability, particularly with reference to the changes in crop patterns. While the farmers have altered their cropping patterns to grow crops that give good yields in the altered climate, they have concerns with the quality of the seed supplied by the government. One of the tribal farmers laments:

> The seed that we are getting from the government is not of good quality. We are now going to Odisha for seed and are using that for farming.

Eventually, farmers believe that there are a series of challenges that cumulatively make farming an unviable occupation and an unpredictable source of livelihood. One of the tribal farmers bemoans:

> Farming is not profitable. But then we are doing this because we do not want to give up on what we have been doing so far. It is generally said that farming can bring in lakhs of rupees in revenue. But that is not true.

Another constraint involves the reclamation of marine lands for construction, thereby effectively reducing the fish drying and boat parking area for the traditional fishermen, one of whom states:

> There is also no place for us to tie our boats. It is also becoming increasingly difficult to dry our nets or sell freshly caught fish by the shore.

CONCLUSION

Connecting scientific research with the adaptation needs of the indigenous communities is of critical significance to avoid the risk of disempowering the actors that these initiatives seek to support (Cochrane & Tamiru, 2016). This is especially so in the context of communities that predominantly comprise climate-susceptible livelihood sectors including fisheries and agriculture and thereby confront more severe constraints on their ability to adapt (Feng et al., 2010; Thornton et al., 2008). From the current research, it can be inferred that fishing and farming communities perceive climate change and have responded through incremental adjustments. However, continuing changes in climate may exceed the capacity of these communities to successfully adapt through incremental adaptation mechanisms. Vulnerable communities like farmers and fisher folk need to essentially locate their adaptation efforts within the framework of transformative adaptation, which attempts to modify the basic attributes of the existing systems as a response to the actual or anticipated effects of change. The transformational strategy presents options that these communities can explore (Pelling et al., 2015) to reorganize institutions and structures when incremental adaptation reaches its limits.

References

Adger, W. N. (2001). Scales of Governance and environmental justice for adaptation and mitigation of climate change. *Journal of International Development, 13*, 921–931.

Agarwal, A. (2009). Local institutions and adaptation to climate change. In M. Robins & N. Andrew (Eds.). *Social dimensions of climate change: Equity and vulnerability in a warming world* (pp. 173–198). The World Bank.

Ayers, J., & Dodman, D. (2010). Climate change adaptation and development I: The state of the debate. *Progress in Development Studies, 10*(2), 161–168.

Braun, V., & Clarke, V. (2006). Using thematic analysis in psychology. *Qualitative Research in Psychology, 3*, 77–101.

Cochrane, L., & Tamiru, Y. (2016). Ethiopia's productive safety net program: Power, politics and practice. *Journal of International Development, 28*(5), 649–665.

Cruikshank, J. (2001). Glaciers and climate change: Perspectives from oral tradition. *Arctic, 54*, 377–393.

Eriksen, S. E. H., & Brown, K. (2011). Sustainable adaptation to climate change. *Climate and Development, 3*, 3–6.

Eriksen, S. E. H., & O'Brien, K. (2007). Vulnerability, poverty and the need for sustainable adaptation measures. *Climate Policy, 7*, 337–352.

Feenstra, J., Burton, I., Smith, J., & Tol, R. (Eds.). (1991). *Handbook on methods for climate impact assessment and adaptation strategies*. Institute for Environmental Studies.

Feng, S., Krueger, A. B., & Oppenheimer, M. (2010). Linkages among climate change crop yields and Mexico-US cross-border migration. *Proceedings of the National Academy of Sciences of the United States of America, 107*(32), 14257–14262.

Fox, S. (2002). There are things that are really happening: Inuit perspectives on the evidence and impacts of climate change in Nunavut. In I. Krupnik & D. Jolly (Eds.), *The earth is faster now: Indigenous observations of arctic environmental change* (pp. 12–53). Arctic Research Consortium of the United States.

Glaser, B. G. (1965). The constant comparative method of qualitative analysis. *Social Problems, 12*(4), 436–445.

Hallberg, L.R.-M. (2009). The "core category" of grounded theory: Making constant comparisons. *International Journal of Qualitative Studies on Health and Well-Being, 1,* 141–148.

IPCC. (2012). *Managing the risks of extreme events and disasters to advance climate change adaptation.* A special report of Working Groups I and II of the intergovernmental panel on climate change. Cambridge University Press.

IPCC. (2014). Climate change 2014: Impacts, adaptation and vulnerability: Part A: *Global and sectoral aspects: Contribution of Working Group II to the fifth assessment report of the intergovernmental panel on climate change: Summary for policy makers.* Cambridge University Press.

Klein, R. J. T. (1998). Towards better understanding, assessment and funding of climate adaptation. *Change, 44,* 15–19.

Klein, R. J. T. (2003). Adaptation to climate variability and change: What is optimal and appropriate? In C. Giupponi & M. Schechter (Eds.), *Climate change in the mediterranean: Socio-economic perspectives of impacts, vulnerability and adaptation.* Edward Elgar.

Lebel, L. (2013). Local knowledge and adaptation to climate change in natural resource-based societies of the Asia-Pacific. *Mitigation and Adaptation Strategies for Global Change, 18*(7), 1057–1076.

Lemos, M. C., Boyd, E., Tompkins, E. L., Osbahr, H., & Liverman, D. (2007). Developing adaptation and adapting development. *Ecology and Society, 12*(2), 26.

Maynard, N. G. (Ed.). (1998). Native peoples—Native homelands climate change workshop. *US global change research program final report* (pp. 93).

McNeeley, S. (2017). Sustainable climate change adaptation in Indian Country. *Weather, Climate and Society, 9,* 393–404.

McNeeley, S. M. (2009). *Seasons out of balance: Climate change impacts, vulnerability and sustainable adaptation in interior Alaska* [Ph.D. Dissertation, University of Alaska].

McNeeley, S. M. (2012). Examining barriers and opportunities for sustainable adaptation to climate change in interior Alaska. *Climatic Change, 111,* 835–857.

McNeeley, S. M., & Shulski, M. D. (2011). An unbroken chain of injustice: The Dawes Act, Native American trusts and Cobell vs. Salazar. *Gonzaga Law Review, 46*, 609–658.

Morchain, D. (2018). Rethinking the framing of climate change adaptation: Knowledge, power and politics. In S. Klepp & L. Chavez-Rodriguez (Eds.), *A critical approach to climate change adaptation: Discourse, policies and practices* (pp. 55–73). Routledge.

Nagoda, S. (2015). New discourses but same old development approaches? Climate change adaptation policies, chronic food insecurity and development interventions in Northwestern Nepal. *Global Environmental Change, 35*, 570–579.

Nuttall, M., Berkes, F., Forbes, B., Kofinas, G., Vlassova, T., & Wenzel, G. (2005). Hunting, herding, fishing and gathering: Indigenous peoples and renewable resource use in the arctic. In C. Symon, L. Arris, & B. Heal (Eds.), *Arctic climate impact assessment* (pp. 649–690). Cambridge University Press.

Orlove, B. (2009). The past, the present and some possible futures of adaptation. In W. N. Adger, I. Lorenzoni, & K. O'Brien (Eds.), *Adapting to climate change: Thresholds, values, governance* (pp. 131–163). Cambridge University Press.

Otto-Banaszak, I., Matczak, P., Wesseler, J., & Wechsung, F. (2011). Different perceptions of adaptation to climate change: A mental model approach applied to the evidence from expert interviews. *Regional Environmental Change, 11*(2), 217–228.

Pelling, M., O'Brien, K., & Matyas, D. (2015). Adaptation and transformation. *Climatic Change, 133*, 113–127.

Risbey, J., Kandlikar, M., Dowlatabadi, H., & Graetz, D. (1999). Scale, context and decision-making in agricultural adaptation to climate variability and change. *Mitigation and Adaptation Strategies for Global Change, 4*(2), 137–165.

Thornton, P. K., et al. (2008). Climate change and poverty in Africa: Mapping hotspots of vulnerability. *African Journal of Agricultural and Resource Economics, 2*(1), 24–44.

Tonmoy, F. N., Rissik, D., & Palutikof, J. P. (2019). A three-tier risk assessment process for climate change adaptation at a local scale. *Climatic Change, 153*, 539–557.

Bibliography—Traditional and Local Knowledge

This compilation of articles has been meticulously curated through an advanced search on EBSCO Discovery Service, utilizing the keywords "Traditional knowledge," "Local Knowledge," "Climate Change," and "Adaptation" with a focus on the Abstract field. The rigorous selection criteria ensure that all included articles are peer-reviewed, guaranteeing their academic reliability and credibility. The disciplines chosen for search reflect a diverse and multidimensional exploration of the intersection between traditional and local knowledge, climate change, and adaptation.

We chose the following disciplines:
Agriculture & Agribusiness
Anthropology
Business & Management
Communication & Mass Media
Earth & Atmospheric Sciences
Economics
Environmental Sciences
Ethnic & Cultural Studies
Forestry
Geography & Cartography
Geology
History
Language & Linguistics
Law
Mining & Mineral Resources

Political Science
Politics & Government
Social Sciences & Humanities
Social Work
Sociology
Women's Studies & Feminism

Subsequently, we looked at the titles of the books/articles and eliminated those entries where the tiles were not fitting into the direct scope of this volume.

Aggarwal, P. K., Jarvis, A., Campbell, B. M., Zougmoré, R. B., Khatri-Chhetri, A., Vermeulen, S. J., Loboguerrero, A. M., Sebastian, L. S., Kinyangi, J., Bonilla-Findji, O., Radeny, M., Recha, J., Martinez-Baron, D., Ramirez-Villegas, J., Huyer, S., Thornton, P., Wollenberg, E., Hansen, J., Alvarez-Toro, P., & Yen, B. T. (2018). The climate-smart village approach: Framework of an integrative strategy for scaling up adaptation options in agriculture. *Ecology and Society, 23*(1), 14–14.

Aguilera, E., Díaz-Gaona, C., García-Laureano, R., Reyes-Palomo, C., Guzmán, G. I., Ortolani, L., Sánchez-Rodríguez, M., & Rodríguez-Estévez, V. (2020). Agroecology for adaptation to climate change and resource depletion in the Mediterranean region. A Review. *Agricultural Systems, 181*, 102809.

Ahmad, W. K., & Ariana, L. (2018). Impact of climate change on indigenous people and adaptive capacity of Bajo Tribe, Indonesia. *Environmental Claims Journal, 30*(4), 302–313.

Akinsemolu, A. A., & Olukoya, O. A. P. (2020). The vulnerability of women to climate change in coastal regions of Nigeria: A case of the Ilaje community in Ondo State. *Journal of Cleaner Production, 246*(2), 119015.

Akinyemi, F. O. (2017). Climate change and variability in semiarid Palapye, Eastern Botswana: An assessment from smallholder farmers' perspective. *Weather, Climate & Society, 9*(3), 349–365.

Alamgir, M., Pretzsch, J., & Turton, S. M. (2014). Climate change effects on community forests: Finding through user's lens and local knowledge. *Small-Scale Forestry, 13*(4), 445–460.

Allen, F. (2023). Nigeria: Decolonial climate adaptation and conflict: Evidence from coastal communities of the Niger Delta. *Conflict Studies Quarterly, 42*, 3–23.

Amoak, D., Luginaah, I., & McBean, G. (2022). Climate change, food security, and health: Harnessing agroecology to build climate-resilient communities. *Sustainability (2071–1050), 14*(21), 13954.

Arku, F. S. (2013). Local creativity for adapting to climate change among rural farmers in the semi-arid region of Ghana. *International Journal of Climate Change Strategies & Management, 5*(4), 418–430.

Arseni, K., Khmelnitskaya, Y., Dugina, M., Borzenko, T., & Tysiachniouk, M. S. (2022). Traditional livelihood, unstable environment: Adaptation of traditional fishing and reindeer herding to environmental change in the Russian Arctic. *Sustainability, 14*(19), 12640–12640.

Asante, F., Guodaar, L., & Arimiyaw, S. (2021). Climate change and variability awareness and livelihood adaptive strategies among smallholder farmers in semi-arid northern Ghana. *Environmental Development, 39*, 100629.

Bahadur, A. V., Ibrahim, M., & Tanner, T. (2013). Characterising resilience: Unpacking the concept for tackling climate change and development. *Climate & Development, 5*(1), 55–65.

Baird, J., Plummer, R., & Pickering, K. (2014). Priming the governance system for climate change adaptation: The application of a social-ecological inventory to engage actors in Niagara, Canada. *Ecology & Society, 19*(1), 245–255.

Barua, P., & Rahman, S. H. (2018). The role of indigenous knowledge and coastal resource management in addressing the climate change impact on Southeastern Bangladesh. *IUP Journal of Knowledge Management, 16*(2), 49–71.

Basel, B., Goby, G., & Johnson, J. (2020). Community-based adaptation to climate change in villages of Western Province, Solomon Islands. *Marine Pollution Bulletin, 156*, 111266.

Basuki, T. M., Nugroho, H. Y. S. H., Indrajaya, Y., Pramono, I. B., Nugroho, N. P., Supangat, A. B., Indrawati, D. R., Savitri, E., Wahyuningrum, N., Purwanto, Cahyono, S. A., Putra, P. B., Adi, R. N., Nugroho, A. W., Auliyani, D., Wuryanta, A., Riyanto, H. D., Harjadi, B., Yudilastyantoro, C., & Hanindityasari, L. (2022). Improvement of integrated watershed management in Indonesia for mitigation and adaptation to climate change: A review. *Sustainability (2071–1050), 14*(16), 9997–9997.

Becken, S., Lama, A. K., & Espiner, S. (2013). The cultural context of climate change impacts: Perceptions among community members in the Annapurna Conservation Area, Nepal. *Environmental Development, 8*, 22–37.

Belay, B. S., & Tebeje, M. (2023). Does local cognition of climate change really matters in adaptation: Farmer perspectives. *Local Environment, 28*(3), 255–276.

Beretić, N., Bauer, A., Funaro, M., Spano, D., & Marras, S. (2023). A participatory framework to evaluate coherence between climate change adaptation and sustainable development policies. *Environmental Policy & Governance, 1–16.*

Berkes, F., & Jolly, D. (2002). Adapting to climate change: Social-ecological resilience in a Canadian Western arctic community. *Conservation Ecology, 5*(2), 18.

Bhatta, L. D., Udas, E., Khan, B., Ajmal, A., Amir, R., & Ranabhat, S. (2020). Local knowledge based perceptions on climate change and its impacts in the

Rakaposhi valley of Gilgit-Baltistan, Pakistan. *International Journal of Climate Change Strategies & Management, 12*(2), 222–237.

Bhawra, J. (2022). Decolonizing digital citizen science: Applying the bridge framework for climate change preparedness and adaptation. *Societies, 12*(2), 71.

Bhowmick, D. (2023). Political ecology of climate change in Sundarbans, India: Understanding well-being, social vulnerabilities, and community perception. *Environmental Quality Management, 1.*

Biggs, E. M., Tompkins, E. L., Allen, J., Moon, C., & Allen, R. (2013). Agricultural adaptation to climate change: Observations from the Mid-Hills of Nepal. *Climate & Development, 5*(2), 165–173.

Blackett, P., FitzHerbert, S., Luttrell, J., Hopmans, T., Lawrence, H., & Colliar, J. (2022). Marae-opoly: Supporting localised Māori climate adaptation decisions with serious games in Aotearoa New Zealand. *Sustainability Science, 17*(2), 415–431.

Bruckmann, L., Tsobgou, L. D., Marcoty, P., & Schmitz, S. (2022). Local perception of climate change and adaptation in the highlands of Cameroon. *African Geographical Review,* 1–16.

Butler, J. R. A., Wise, R. M., Meharg, S., Peterson, N., Bohensky, E. L., Lipsett-Moore, G., Skewes, T. D., Hayes, D., Fischer, M., & Dunstan, P. (2022). Walking along with development: Climate resilient pathways for political resource curses. *Environmental Science & Policy, 128,* 228–241.

Chisale, H. L. W., Chirwa, P. W., & Babalola, F. D. (2023). Awareness, knowledge and perception of forest dependent communities on climate change in Malawi: A case of Mchinji and Phirilongwe forest reserves in Malawi. *Journal of Sustainable Forestry, 42*(7), 728–745.

Chowdhury, R. B., & Moore, G. A. (2017). Floating agriculture: A potential cleaner production technique for climate change adaptation and sustainable community development in Bangladesh. *Journal of Cleaner Production, 150,* 371–389.

Clissold, R., McNamara, K. E., Westoby, R., & Wichman, V. (2023). Experiencing and responding to extreme weather: Lessons from the Cook Islands. *Local Environment, 28*(5), 645–661.

Cottrell, C. (2022). Avoiding a new era in biopiracy: Including indigenous and local knowledge in nature-based solutions to climate change. *Environmental Science & Policy, 135,* 162–168.

Cuaton, G. P., & Su, Y. (2023). Potentials and pitfalls of social capital ties to climate change adaptation: An exploratory study of indigenous peoples in the Philippines. *Third World Quarterly, 44*(7), 1565–1585.

Dannevig, H., & Aall, C. (2015). The regional level as boundary organization? An analysis of climate change adaptation governance in Norway. *Environmental Science & Policy, 54,* 168–175.

Das, S., & Mishra, A. J. (2023a). Climate change and the Western Himalayan community: Exploring the local perspective through food choices. *AMBIO - A Journal of the Human Environment, 52*(3), 534–545.

Das, S., & Mishra, A. J. (2023b). Climate change, dietary shift, and traditional norms in the western Himalayan region, India. *Climate & Development, 15*(6), 509–517.

de Carvalho, D. A., Amaral, S., & Alves, L. M. (2023). Climate change adaptation frameworks in fishing communities: A systematic review. *Ocean & Coastal Management, 243*, 106754.

de Scally, D., & Doberstein, B. (2022). Local knowledge in climate change adaptation in the Cook Islands. *Climate & Development, 14*(4), 360–373.

Derbile, E. K., & File, D. J. M. (2016). Community risk assessment of rainfall variability under rain-fed agriculture: The potential role of local knowledge in Ghana. *Ghana Journal of Development Studies, 13*(2), 66–83.

Donghyun, K., & Jung, E. K. (2020). Building consensus with local residents in community-based adaptation planning: The case of Bansong Pilbongoreum community in Busan, South Korea. *Sustainability, 12*(4), 1559–1559.

Dramani, J. M. F., Jarawura, F. X., & Derbile, E. K. (2023). Adapting to climate change: Perspectives from smallholder farmers in North-western Ghana. *Cogent Social Sciences, 9*(1), 2228064.

Dutra, L. X. C., Haywood, M. D. E., Singh, S., Ferreira, M., Johnson, J. E., Veitayaki, J., Kininmonth, S., Morris, C. W., & Piovano, S. (2021). Synergies between local and climate-driven impacts on coral reefs in the Tropical Pacific: A review of issues and adaptation opportunities. *Marine Pollution Bulletin, 164*, 111922.

Eitzel, M. V., Solera, J., Wilson, K. B., Neves, K., Fisher, A. C., Veski, A., Omoju, O. E., Ndlovu, A. M., & Hove, E. M. (2020a). Indigenous climate adaptation sovereignty in a Zimbabwean agro-pastoral system: Exploring definitions of sustainability success using a participatory agent-based model. *Ecology & Society, 25*(4), 1–46.

Endalamaw, T. B., & Darr, D. (2021). Institutional and technological innovation for the bamboo sector as an instrument for development and climate change resilience in Ethiopia. *African Journal of Science, Technology, Innovation & Development, 13*(7), 817–828.

Everlyne, B., Obwocha, J., Ramisch, J., Duguma, L., & Orero, L. (2022). The relationship between climate change, variability, and food security: understanding the impacts and building resilient food systems in West Pokot County, Kenya. *Sustainability, 14*(2), 765–765.

Fatorić, S., & Morén-Alegret, R. (2013). Integrating local knowledge and perception for assessing vulnerability to climate change in economically dynamic coastal areas: The case of natural protected area Aiguamolls de l'Empordà, Spain. *Ocean & Coastal Management, 85*, 90–102.

Filho, W. L., Wolf, F., Totin, E., Zvobgo, L., Simpson, N. P., Musiyiwa, K., Kalangu, J. W., Sanni, M., Adelekan, I., Efitre, J., Donkor, F. K., Balogun, A., Mucova, S. A. R., & Ayal, D. Y. (2023). Is indigenous knowledge serving climate adaptation? Evidence from various African regions. *Development Policy Review, 41*(2), 1–22.

Filho, L. W., Barbir, J., Gwenzi, J., Ayal, D., Simpson, N. P., Adeleke, L., Tilahun, B., Chirisa, I., Gbedemah, S. F., Nzengya, D. M., Sharifi, A., Theodory, T., & Yaffa, S. (2022). The role of indigenous knowledge in climate change adaptation in Africa. *Environmental Science & Policy, 136*, 250–260.

Flynn, M., Ford, J. D., Pearce, T., & Harper, S. L. (2018). Participatory scenario planning and climate change impacts, adaptation and vulnerability research in the Arctic. *Environmental Science & Policy, 79*, 45–53.

Ford, J. D., Willox, C. A., Chatwood, S., Furgal, C., Harper, S., Mauro, I., & Pearce, T. (2014). Adapting to the effects of climate change on Inuit health. *American Journal of Public Health, 104*(S3), e9–e17.

Ford, J. D., Pearce, T., Canosa, I. V., & Harper, S. (2021). The rapidly changing Arctic and its societal implications. *WIREs: Climate Change, 12*(6), 1–27.

Frazier, T. G., Wood, N., & Yarnal, B. (2010). Stakeholder perspectives on land-use strategies for adapting to climate-change-enhanced coastal hazards: Sarasota, Florida. *Applied Geography, 30*(4), 506–517.

Fu, Y., Grumbine, R., Wilkes, A., Wang, Y., Xu, J. C., & Yang, Y. P. (2012). Climate change adaptation among Tibetan pastoralists: Challenges in enhancing local adaptation through policy support. *Environmental Management, 50*(4), 607–621.

Gachathi, F. N., & Eriksen, S. (2011). Gums and resins: The potential for supporting sustainable adaptation in Kenya's drylands. *Climate & Development, 3*(1), 59–70.

Gagné, K. (2016). Cultivating ice over time: On the idea of timeless knowledge and places in the Himalayas. *Anthropologica, 58*(2), 193–210.

Gaisie, E., & Cobbinah, P. B. (2023). Planning for context-based climate adaptation: Flood management inquiry in Accra. *Environmental Science & Policy, 141*, 97–108.

Galappaththi, E. K., & Schlingmann, A. (2023). The sustainability assessment of indigenous and local knowledge-based climate adaptation responses in agricultural and aquatic food systems. *Current Opinion in Environmental Sustainability, 62*, 101276.

Galappaththi, E. K., Ford, J. D., & Bennett, E. M. (2019a). A framework for assessing community adaptation to climate change in a fisheries context. *Environmental Science & Policy, 92*, 17–26.

Galappaththi, E. K., Ford, J. D., Bennett, E. M., & Berkes, F. (2019b). Climate change and community fisheries in the Arctic: A case study from Pangnirtung, Canada. *Journal of Environmental Management, 250,* 109534.

Galappaththi, E. K., Susarla, V. B., Loutet, S. J. T., Ichien, S. T., Hyman, A. A., & Ford, J. D. (2022). Climate change adaptation in fisheries. *Fish & Fisheries, 23*(1), 4–21.

Germano, M. (2022). "Neutral" representations of Pacific Islands in the IPCC special report of 1.5°C global warming. *Australian Geographer, 53*(1), 23–39.

Ghimire, R., & Chhetri, N. (2022). Challenges and prospects of local adaptation plans of action (LAPA) initiative in Nepal as everyday adaptation. *Ecology and Society, 27*(4), 28–28.

Ghimire, R., & Chhetri, N. (2023). Coproductive imaginaries for climate change adaptation: A case of adaptation initiatives in the Gandaki river basin, Western Nepal. *Professional Geographer, 75*(2), 324–334.

Gianelli, I., Ortega, L., Pittman, J., Vasconcellos, M., & Defeo, O. (2021). Harnessing scientific and local knowledge to face climate change in small-scale fisheries. *Global Environmental Change Part a: Human & Policy Dimensions, 68,* 102253.

Githiora, Y. W., Owuor, M. A., Abila, R., Oriaso, S., & Olago, D. O. (2023). Perceptions, trends and adaptation to climate change in Yala wetland, Kenya. *International Journal of Climate Change Strategies & Management, 15*(5), 690–711.

Goeldner-Gianella, L., Grancher, D., Magnan, A. K., de Belizal, E., & Duvat, V. K. E. (2019). The perception of climate-related coastal risks and environmental changes on the Rangiroa and Tikehau atolls, French Polynesia: The role of sensitive and intellectual drivers. *Ocean & Coastal Management, 172,* 14–29.

Granderson, A. A. (2017). The role of traditional knowledge in building adaptive capacity for climate change: Perspectives from Vanuatu. *Weather, Climate & Society, 9*(3), 545–561.

Greene, C., Wilmer, H., Ferguson, D. B., Crimmins, M. A., & McClaran, M. P. (2022). Using scale and human agency to frame ranchers' discussions about socio-ecological change and resilience. *Journal of Rural Studies, 96,* 217–226.

Griffin, C., Wreford, A., & Cradock-Henry, N. A. (2023). "As a farmer you've just got to learn to cope": Understanding dairy farmers' perceptions of climate change and adaptation decisions in the lower south Island of Aotearoa-New Zealand. *Journal of Rural Studies, 98,* 147–158.

Guoping, W., Lun, Y., Moucheng, L., Zhidong, L., Siyuan, H., & Qingwen, M. (2021). The role of local knowledge in the risk management of extreme climates in local communities: A case study in a nomadic NIAHS Site. *Journal of Resources & Ecology, 12*(4), 532–542.

Hallberg-Sramek, I., Reimerson, E., Priebe, J., Nordström, E.-M., Mårald, E., Sandström, C., & Nordin, A. (2022). Bringing "Climate-Smart Forestry" down to the local level—Identifying barriers, pathways and indicators for its implementation in practice. *Forests, 13*(1), 98.

Holzkämper, A. (2017). Adapting agricultural production systems to climate change—What's the use of models? *Agriculture; Basel, 7*(10), 86.

Hunter, N. B., North, M. A., Roberts, D. C., & Slotow, R. (2020). A systematic map of responses to climate impacts in urban Africa. *Environmental Research Letters, 15*(10), 103005–103005.

Inaotombi, S., & Mahanta, P. C. (2019). Pathways of socio-ecological resilience to climate change for fisheries through indigenous knowledge. *Human & Ecological Risk Assessment, 25*(8), 2032–2044.

Iocca, L., & Fidélis, T. (2022). Traditional communities, territories and climate change in the literature—Case studies and the role of law. *Climate & Development, 14*(6), 537–556.

Irisha, J. I. J., & Esa, N. (2018). Contribution of local knowledge towards urban agroforestry as a sustainable approach on climate change adaptation. *SHS Web of Conferences, 45*, 03001.

Iwama, A. Y., Araos, F., Anbleyth-Evans, J., Marchezini, V., Ruiz-Luna, A., Ther-Ríos, F., Bacigalupe, G., & Perkins, P. E. (2021). Multiple knowledge systems and participatory actions in slow-onset effects of climate change: Insights and perspectives in Latin America and the Caribbean. *Current Opinion in Environmental Sustainability, 50*, 31–42.

Janif, S. Z., Nunn, P. D., Geraghty, P., Aalbersberg, W., Thomas, F. R., & Camailakeba, M. (2016). Value of traditional oral narratives in building climate-change resilience: Insights from rural communities in Fiji. *Ecology & Society, 21*(2), 667–676.

Jellason, N. P., Conway, J. S., Baines, R. N., & Ogbaga, C. C. (2021). A review of farming challenges and resilience management in the Sudano-Sahelian drylands of Nigeria in an era of climate change. *Journal of Arid Environments, 186*, 104398.

Jigyasu, R. (2020). Managing cultural heritage in the face of climate change. *Journal of International Affairs, 73*(1), 87–100.

Jonsson, A. C., & Lundgren, L. (2015). Vulnerability and adaptation to heat in cities: Perspectives and perceptions of local adaptation decision-makers in Sweden. *Local Environment, 20*(4), 442–458.

Kaptijn, E. (2018). Learning from ancient water management: Archeology's role in modern-day climate change adaptations. *WIRES Water, 5*(1), e.1256.

Kassie, B., Hengsdijk, H., Rötter, R., Kahiluoto, H., Asseng, S., & Ittersum, M. (2013). Adapting to climate variability and change: Experiences from cereal-based farming in the Central Rift and Kobo valleys, Ethiopia. *Environmental Management, 52*(5), 1115–1131.

Karume, K., Mondo, J. M., Chuma, G. B., Ibanda, A., Bagula, E. M., Aleke, A. L., Ndjadi, S., Ndusha, B., Ciza, P. A., Cizungu, N. C., Muhindo, D., Egeru, A., Nakayiwa, F. M., Majaliwa, J. G. M., Mushagalusa, G. N., & Ayagirwe, R. B. B. (2022). Current practices and prospects of climate-smart agriculture in Democratic Republic of Congo: A review. *Land, 11*(10), 1850–1850.

Kerr, E. (2022). "We are not drowning, we are fighting": A critical examination of the climate change adaptation law and policy framework in the Pacific Islands. *Te Mata Koi: Auckland University Law Review, 28*, 47–77.

Khalil, M. B., Jacobs, B. C., McKenna, K., & Kuruppu, N. (2020). Female contribution to grassroots innovation for climate change adaptation in Bangladesh. *Climate & Development, 12*(7), 664–676.

Kieslinger, J., Pohle, P., Buitrón, V., & Peters, T. (2019). Encounters between experiences and measurements: The role of local knowledge in climate change research. *Mountain Research & Development, 39*(2), R55–R68.

Kim, D., & Kang, J. E. (2020). Building consensus with local residents in community-based adaptation planning: The case of Bansong Pilbongoreum Community in Busan, South Korea. *Sustainability (2071–1050), 12*(4), 1559.

Klein, J. A., Hopping, K. A., Yeh, E. T., Nyima, Y., Boone, R. B., & Galvin, K. A. (2014). Unexpected climate impacts on the Tibetan Plateau: Local and scientific knowledge in findings of delayed summer. *Global Environmental Change Part a: Human & Policy Dimensions, 28*, 141–152.

Klenk, N., Fiume, A., Meehan, K., & Gibbes, C. (2017). Local knowledge in climate adaptation research: Moving knowledge frameworks from extraction to co-production. *WIREs: Climate Change, 8*(5), e475.

Klöck, C. (2019). Dealing with climate change in the German Wadden Sea: Perceptions, measures, and contestation on Hallig Hooge. *Ocean & Coastal Management, 179*, 104864.

Konnov, A., Khmelnitskaya, Y., Dugina, M., Borzenko, T., & Tysiachniouk, M. S. (2022). Traditional livelihood, unstable environment: Adaptation of traditional fishing and reindeer herding to environmental change in the Russian Arctic. *Sustainability, 14*(19), 12640–12640.

Korovulavula, I., Nunn, P. D., Kumar, R., & Fong, T. (2020). Peripherality as key to understanding opportunities and needs for effective and sustainable climate-change adaptation: A case study from Viti Levu Island, Fiji. *Climate & Development, 12*(10), 888–898.

Leon, J. X., Hardcastle, J., James, R., Albert, S., Kereseka, J., & Woodroffe, C. D. (2015). Supporting local and traditional knowledge with science for adaptation to climate change: Lessons learned from participatory three-dimensional modeling in BoeBoe, Solomon Islands. *Coastal Management, 43*(4), 424–438.

Iocca, L., & Fidélis, T. (2023). Is there a place for indigenous peoples and local communities in climate change policy and governance? Learnings from a Brazilian case. *Land, 12*(9), 1647–1647.

Eitzel, M. V., Solera, J. K. B., Neves, W. K., Fisher, A. C., Veski, A., Omoju, O. E., Ndlovu, A. M., & Hove, E. M. (2020b). Indigenous climate adaptation sovereignty in a Zimbabwean agro-pastoral system: Exploring definitions of sustainability success using a participatory agent-based model. *Ecology and Society, 25*(4), 13–13.

Makondo, C. C., & Thomas, D. S. G. (2018). Climate change adaptation: Linking indigenous knowledge with western science for effective adaptation. *Environmental Science & Policy, 88*, 83–91.

Malherbe, W. S., Aswani, S., Lemahieu, A., Scott, L., Mahatante, P. T., & Randrianarimanana, J. V. (2018). Local perceptions of environmental changes in fishing communities of southwest Madagascar. *Ocean & Coastal Management, 163*, 209–221.

Manh, N. T., & Ahmad, M. M. (2021). Indigenous farmers' perception of climate change and the use of local knowledge to adapt to climate variability: A case study of Vietnam. *Journal of International Development, 33*(7), 1189–1212.

Marshall, N., Dowd, A. M., Fleming, A., Gambley, C., Howden, M., Jakku, E., Larsen, C., Marshall, P., Moon, K., Park, S., & Thorburn, P. (2014). Transformational capacity in Australian peanut farmers for better climate adaptation. *Agronomy for Sustainable Development, 34*(3), 583–591.

Marzano, M. (2006). Changes in the weather: A Sri Lankan village case study. *Anthropology in Action, 13*(3), 63–76.

Masud-All-Kamal, M., & Nursey-Bray, M. (2022). Best intentions and local realities: Unseating assumptions about implementing planned community-based adaptation in Bangladesh. *Climate & Development, 14*(9), 794–803.

Mathew, S., Trück, S., & Henderson-Sellers, A. (2012). Kochi, India case study of climate adaptation to floods: Ranking local government investment options. *Global Environmental Change Part a: Human & Policy Dimensions, 22*(1), 308–319.

McNamara, K. E., & Buggy, L. (2017). Community-based climate change adaptation: A review of academic literature. *Local Environment, 22*(4), 443–460.

Metternicht, G., Sabelli, A., & Spensley, J. (2014). Climate change vulnerability, impact and adaptation assessment lessons from Latin America. *International Journal of Climate Change Strategies & Management, 6*(4), 442–476.

Morgan, R. A. (2013). Histories for an uncertain future: Environmental history and climate change. *Australian Historical Studies, 44*(3), 350–360.

Naess, L. O. (2013). The role of local knowledge in adaptation to climate change. *WIREs: Climate Change, 4*(2), 99–106.

Nasir, M. J., Khan, A. S., & Alam, S. (2018). Climate change and agriculture: An overview of farmers perception and adaptations in Balambat Tehsil, district Dir Lower, Pakistan. *Sarhad Journal of Agriculture, 34*(1), 85–92.

Naznin, A. (2013). The knowledge: Climate change policy versus "Char" people: An Anthropological framework. *International Journal of Climate Change: Impacts & Responses, 4*(3), 13–30.

Nursey-Bray, M., Palmer, R., Stuart, A., Arbon, V., & Rigney, L. I. (2020). Scale, colonisation and adapting to climate change: Insights from the Arabana people, South Australia. *Geoforum, 114*, 138–150.

O'Donnell, C., Recharte, J., & Taber, A. (2016). Climate change, mountain people and water resources—The experiences of the Mountain Institute, Peru. *Unasylva, 67*(246), 75–80.

Ogra, M., Manral, U., Platt, R. V., Badola, R., & Butcher, L. (2020). Local perceptions of change in climate and agroecosystems in the Indian Himalayas: A case study of the Kedarnath Wildlife Sanctuary (KWS) landscape, India. *Applied Geography, 125*, 102339.

Ogunyiola, A., Gardezi, M., & Vij, S. (2022). Smallholder farmers' engagement with climate smart agriculture in Africa: Role of local knowledge and upscaling. *Climate Policy, 22*(4), 411–426.

Ojoyi, M. M., & Kahinda, J. M. M. (2015). An analysis of climatic impacts and adaptation strategies in Tanzania. *International Journal of Climate Change Strategies & Management, 7*(1), 97–115.

Ogalleh, S. A., Vogl, C. R., Eitzinger, J., & Hauser, M. (2012a). Local perceptions and responses to climate change and variability: The case of Laikipia District, Kenya. *Sustainability, 4*(12), 3302–3325.

Ogalleh, A. S., Vogl, C., & Hauser, M. (2012b). Reading from farmers' scripts: Local perceptions of climate variability and adaptations in Laikipia, Rift Valley, Kenya. *Journal of Agriculture, Food Systems & Community Development, 3*(2), 77–94.

Pauli, N., Williams, M., Henningsen, S., Davies, K., Chhom, C., van Ogtrop, F., Hak, S., Boruff, B., & Neef, A. (2021). "Listening to the sounds of the water": Bringing together local knowledge and biophysical data to understand climate-related hazard dynamics. *International Journal of Disaster Risk Science, 12*(3), 326–340.

Peñalba, E. H., David, A. P. J., Mabanta, M. J. D., Samaniego, C. R. C. & Ellamil, S. D. S. (2021). Climate change adaptation: The case of coastal communities in the philippines. *Journal of the Geographical Institute "Jovan Cvijic" SASA, 71*(2), 115–133.

Pérez, F. B., & Tomaselli, A. (2021). Indigenous peoples and climate-induced relocation in Latin America and the Caribbean: Managed retreat as a tool or a threat? *Journal of Environmental Studies & Sciences, 11*(3), 352–364.

Petheram, L., Stacey, N., & Fleming, A. (2015). Future sea changes: Indigenous women's preferences for adaptation to climate change on South Goulburn Island, Northern Territory (Australia). *Climate & Development*, 7(4), 339–352.

Petrescu-Mag, R. M., Petrescu, D. C., Muntean, O. L., Petrescu-Mag, I. V., Radu Tenter, A., & Azadi, H. (2022). The nexus of traditional knowledge and climate change adaptation: Romanian farmers' behavior towards landraces. *Local Environment*, 27(2), 229–250.

Phong, T. N., Quang, H. N., & Sang, V. T. (2022). Shoreline change and community-based climate change adaptation: Lessons learnt from Brebes Regency, Indonesia. *Ocean & Coastal Management*, 216, 106037.

Pinho, P. F., Canova, M. T., Toledo, P. M., Gonzalez, A., Lapola, D. M., Ometto, J. P., & Smith, M. S. (2022). Climate change affects us in the tropics: Local perspectives on ecosystem services and well-being sensitivity in Southeast Brazil. *Regional Environmental Change*, 22(3), 1–17.

Popovici, R., de L. Moraes, A. G., Zhao, M., Zanotti, L., Cherkauer, K. A., Erwin, A. E., Mazer, K. E., Delgado, B. E. F., Cáceres, J. P. P., Ranjan, P., & Prokopy, L. S. (2021). How do indigenous and local knowledge systems respond to climate change? *Ecology & Society*, 26(3), 217–230.

Postigo, J. C. (2021). The role of social institutions in indigenous Andean Pastoralists' adaptation to climate-related water hazards. *Climate & Development*, 13(9), 780–791.

Raihan, F., & Hossain, M. M. (2021). Livelihood vulnerability assessments and adaptation strategies to climate change: A case study in Tanguar haor, Sylhet. *Journal of Water and Climate Change*, 12(7), 3448–3463.

Rajeev, M. M. (2022). The impact of climate change on water resources: Lessons from villages of Tonk District of Rajasthan, India. *Journal of Climate Change*, 8(4), 35–42.

Reid, M. G., Hamilton, C., Reid, S. K., Trousdale, W., Hill, C., Turner, N., Picard, C. R., Lamontagne, C., & Matthews, H. D. (2014). Indigenous climate change adaptation planning using a values-focused approach: A case study with the Gitga'at Nation. *Journal of Ethnobiology*, 34(3), 401–424.

Ricart, S., Castelletti, A., & Gandolfi, C. (2022). On farmers' perceptions of climate change and its nexus with climate data and adaptive capacity. A comprehensive review. *Environmental Research Letters*, 17(8), 083002.

Rijal, S., Gentle, P., Khanal, U., Wilson, C., & Rimal, B. (2022). A systematic review of Nepalese farmers' climate change adaptation strategies. *Climate Policy*, 22(1), 132–146.

Rivera-Ferre, M. G., Di Masso, M., Vara, I., Cuellar, M., Calle, A., Mailhos, M., López-i-Gelats, F., Bhatta, G., & Gallar, D. (2016). Local agriculture traditional knowledge to ensure food availability in a changing climate: Revisiting water management practices in the Indo-Gangetic Plains. *Agroecology & Sustainable Food Systems, 40*(9), 965–987.

Riznic, D., Nikolic, R., & Stojanovic, G. (2014). The economics of climate change and managing the risk caused by the climatic changes at local level. *EMIT: Economics, Management, Information, Technology, 3*(2), 74–86.

Rodríguez, A. G., & Meza, L. E. (2017). Building cooperation agendas from policy dialogue on agriculture and climate change in Latin America and the Caribbean. *Climate & Development, 9*(6), 571–574.

Romero, M. D., Corral, S., & Guimarães, P. Å. (2018). Climate-related displacements of coastal communities in the Arctic: Engaging traditional knowledge in adaptation strategies and policies. *Environmental Science & Policy, 85*, 90–100.

Roös, P. B. (2015). Indigenous knowledge and climate change: Settlement patterns of the past to adaptation of the future. *International Journal of Climate Change: Impacts & Responses, 7*(1), 13–31.

Rushton, B., & Rutty, M. (2023). Gaining insight from the most challenging expedition: Climate change from the perspective of Canadian mountain guides. *Current Issues in Tourism, 26*(23), 3903–3915.

Salgotra, R. K., Zargar, S. M., Sharma, M., & Sood, M. (2018). Traditional knowledge: A therapeutic potential in the scenario of climate change for sustainable development. *Development, 61*(1–4), 140–148.

Sampson, Y., Appiah, D. O., & Siaw, L. P. (2019). Smallholder farmers' perceptions and adaptive response to climate variability and climate change in southern rural Ghana. *Cogent Social Sciences, 5*(1), 1646626.

Sanganyado, E., Teta, C., & Masiri, B. (2018). Impact of African traditional worldviews on climate change adaptation. *Integrated Environmental Assessment & Management, 14*(2), 189–193.

Savage, A., Schubert, L., Huber, C., Bambrick, H., & Hall, N. (2020). Adaptation to the climate crisis: Opportunities for food and nutrition security and health in a Pacific small island state. *Weather, Climate & Society, 12*(4), 745–758.

Schmidt, L., Delicado, A., Gomes, C., Granjo, P., Guerreiro, S., Horta, A., Mourato, J., Prista, P., Saraiva, T., Truninger, M., O'Riordan, T., Santos, F. D., & Penha-Lopes, G. (2013). Change in the way we live and plan the coast: Stakeholders discussions on future scenarios and adaptation strategies. *Journal of Coastal Research, 65*, 1033–1038.

Shaffril, M. H. A., Ahmad, N., Samsuddin, S. F., Samah, A. A., & Hamdan, M. E. (2020). Systematic literature review on adaptation towards climate change impacts among indigenous people in the Asia Pacific regions. *Journal of Cleaner Production, 258*, 120595.

Sharma, R., Jagtap, S., & Rao, P. (2022). Understanding Maharashtra coastal community's perceptions and livelihood resilience to climate change using the community participatory approach. *International Journal of Climate Change: Impacts & Responses, 14*(2), 1–19.

Sidorova, E., & Virla, L. D. (2022). Community-based environmental monitoring (CBEM) for meaningful incorporation of indigenous and local knowledge within the context of the Canadian Northern Corridor Program. *School of Public Policy Publications, 15*(15), 1–50.

Singh, R. K., Singh, A., Zander, K. K., Mathew, S., & Kumar, A. (2021). Measuring successful processes of knowledge co-production for managing climate change and associated environmental stressors: Adaptation policies and practices to support Indian farmers. *Journal of Environmental Management, 282*, 111679.

Stensrud, A. B. (2016). Climate change, water practices and relational worlds in the Andes. *Ethnos: Journal of Anthropology, 81*(1), 75–98.

Sultana, P., & Thompson, P. M. (2017). Adaptation or conflict? Responses to climate change in water management in Bangladesh. *Environmental Science & Policy, 78*, 149–156.

Sultana, T., Islam, M. R., Mustafa, F. B., & Sim, J. O. L. (2022). A systematic review of coastal community adaptation practices in response to climate change-induced tidal inundation. *Journal of Coastal Conservation, 26*(4), 1–16.

Swe, L. M. M., Shrestha, R. P., Ebbers, T., & Jourdain, D. (2015). Farmers' perception of and adaptation to climate-change impacts in the Dry Zone of Myanmar. *Climate & Development, 7*(5), 437–453.

Timko, J., Green, S., Sharples, R., & Grinde, A. (2015). Using a community-driven approach to identify local forest and climate change priorities in Teslin, Yukon. *Cogent Social Sciences, 1*(1), 1047564.

Trundle, A. (2020). Resilient cities in a Sea of Islands: Informality and climate change in the South Pacific. *Cities, 97*, 102496.

Tschanz, L., Arlot, M. P., Philippe, F., Vidaud, L., Morin, S., Maldonado, E., George, E., & Spiegelberger, T. (2022). A transdisciplinary method, knowledge model and management framework for climate change adaptation in mountain areas applied in the Vercors, France. *Regional Environmental Change, 22*(1), 1–15.

Tubridy, F., Lennon, M., & Scott, M. (2022). Managed retreat and coastal climate change adaptation: The environmental justice implications and value of a coproduction approach. *Land Use Policy, 114*, 105960.

Valdivia, C., Seth, A., Gilles, J. L., García, M., Jiménez, E., Cusicanqui, J., Navia, F., & Yucra, E. (2010). Adapting to climate change in Andean ecosystems: Landscapes, capitals, and perceptions shaping rural livelihood strategies and linking knowledge systems. *Annals of the Association of American Geographers, 100*(4), 818–834.

van Gevelt, T., Abok, H., Bennett, M. M., Fam, S. D., George, F., Kulathuramaiyer, N., Low, C. T., & Zaman, T. (2019). Indigenous perceptions of climate anomalies in Malaysian Borneo. *Global Environmental Change Part a: Human & Policy Dimensions, 58*, 1–11.

Veeran, Y., Bose, J. R. S., & Kandasamy, S. (2022). Local knowledge of coastal population to sea level rise and climate change—A case study in fishermen community, Kanyakumari District, Tamil Nadu, India. *Journal of Climate Change, 8*(2), 23–34.

Vignola, R., Harvey, C. A., Bautista-Solis, P., Avelino, J., Rapidel, B., Donatti, C., & Martinez, R. (2015). Ecosystem-based adaptation for smallholder farmers: Definitions, opportunities and constraints. *Agriculture, Ecosystems & Environment, 211*, 126–132.

Vijhani, A., Sinha, V. S. P., Vishwakarma, C. A., Singh, P., Pandey, A., & Govindan, M. (2023). Study of stakeholders' perceptions of climate change and its impact on mountain communities in central Himalaya, India. *Environmental Development, 46*, 100824.

Webler, T., Tuler, S., Dow, K., Whitehead, J., & Kettle, N. (2016). Design and evaluation of a local analytic-deliberative process for climate adaptation planning. *Local Environment, 21*(2), 166–188.

Westoby, R., Clissold, R., McNamara, K. E., Ahmed, I., Resurrección, B. P., Fernando, N., & Huq, S. (2021). Locally led adaptation: Drivers for appropriate grassroots initiatives. *Local Environment, 26*(2), 313–319.

Wilson, A. M. W., & Forsyth, C. (2018). Restoring near-shore marine ecosystems to enhance climate security for island ocean states: Aligning international processes and local practices. *Marine Policy, 93*, 284–294.

Wollenberg, E., Hansen, J., Aggarwal, P. K., Khatri-Chhetri, A., Jarvis, A., Loboguerrero, A. M., Bonilla-Findji, O., Martinez-Baron, D., Campbell, B. M., Zougmoré, R. B., Ouedraogo, M., Vermeulen, S. J., Sebastian, L. S., Yen, B. T., Kinyangi, J., Radeny, M., Recha, J., Thornton, P., Huyer, S., & Ramirez-Villegas, J. (2018). The climate-smart village approach: Framework of an integrative strategy for scaling up adaptation options in agriculture. *Ecology & Society, 23*(1), 474–488.

Yeleliere, E., Antwi-Agyei, P., & Nyamekye, A. B. (2023). Mainstreaming indigenous knowledge systems and practices in climate-sensitive policies for resilient agricultural systems in Ghana. *Society & Natural Resources, 36*(6), 639–659.

Zin, W. Y. L., Teartisup, P., & Kerdseub, P. (2019). Evaluating traditional knowledge on climate change (TKCC): A case study in the Central Dry Zone of Myanmar. *Environment & Natural Resources Journal, 17*(2), 1–29.

Zvobgo, L., Johnston, P., Olagbegi, O. M., Simpson, N. P., & Trisos, C. H. (2023). Role of indigenous and local knowledge in seasonal forecasts and climate adaptation: A case study of smallholder farmers in Chiredzi, Zimbabwe. *Environmental Science & Policy, 145*, 13–28.

Zvobgo, L., Johnston, P., Williams, P. A., Trisos, C. H., & Simpson, N. P. (2022). The role of indigenous knowledge and local knowledge in water sector adaptation to climate change in Africa: A structured assessment. *Sustainability Science, 17*(5), 2077–2092.

INDEX